A-Z
CHEMISTRY

A-Z CHEMISTRY

Prof. Naresh Purohit

CENTRUM PRESS
NEW DELHI-110002 (INDIA)

CENTRUM PRESS
H.O.: 4360/4, Ansari Road, Daryaganj,
New Delhi-110 002 (India)
Ph.: 23278000, 23261597
B.O.: No. 1015, Ist Main Road, BSK IIIrd Stage
IIIrd Phase, IIIrd Block,
Bangalore - 560 085 (India)
Tel.: 080-41723429
Visit us at: www.centrumpress.com

A-Z Chemistry

First Edition, 2009
ISBN 978-93-80106-49-6

PRINTED IN INDIA

Printed at Salasar Imaging Systems, Delhi-110035 (India)

Contents

Preface

Chemistry is the application of the principles and methods of physics and math to chemistry. It can also be regarded as the study of the physical principles underlying chemistry. We want to know how and why materials behave as they do. The ultimate goal of chemistry is to provide a (mathematical) model for all of chemistry.

The purpose of the present work is primarily that of considering the development of certain branches of chemical science. The method of treatment, namely that of selecting particular topics and allocating them to separate chapters—that is, division according to subject-matter and not according to periods of time—has again been adopted. It may be added—as is indicated by the title—that somewhat greater emphasis has been given to certain of the older parts of the science. It covers virtually all topics suggested for the core inorganic requirement for a professional chemistry degree by the Indian Chemical Society Division of Inorganic Chemistry Ad-hoc Subcommittee.

Although the book is intended for the serious student of chemistry, the style of treatment is 'elementary' in the sense that it should appeal to others whose acquaintance with the subject is more limited, and whose chief interests may lie in altogether different kinds of study. It will doubtless be appreciated that in a book of this kind the sketches of individual topics must of necessity be somewhat impressionistic.

Author

Chapter 1

Introductory Ideas

MACROSCOPIC, MICROSCOPIC, AND MOLECULAR VIEWPOINTS

The scope of scientific inquiry is magnificent. It ranges from the very small (the size of an elementary particle, 10^{-16} meters) to the very large (the size of the universe, 10^{26} meters); the very brief (the time required for a photon of light to traverse an atomic nucleus, about 10^{-24} seconds) to the very lengthy (the age of the universe, about 15 billion years, or 10^{18} seconds); the very light (the mass of an electron, 10^{-27} grams) to the very heavy (the mass of all of the matter in the universe, presently unknown, but certainly larger than 10^{58} g).

Thus, scientists must be adept at discussing and comprehending different scales of size, time, mass, energy, electric charge, and so on. Chemists and physicists in particular use words that give some indication of the scale in which they are thinking. Several such words are macroscopic, microscopic, and atomic (molecular). A macroscopic phenomenon is an event or property that is characteristic of an ensemble (i.e., a large collection) of atoms or molecules. It can be observed in a sample of matter of large size, where large implies that the sample can be weighed.

The best available analytical balances are capable of detecting a mass of 1 microgram (10^{-6} g). A microscopic phenomenon is one that can be viewed with an optical microscope, an instrument that allows magnification factors as high as 1000. An atomic (or molecular) phenomenon is an event or property that is characteristic of an individual atom or molecule, or of a small collection of atoms or molecules. Such small samples of matter are too small to be weighed, and too small to be seen with an optical microscopic. Macroscopic properties represent averages or cumulative totals of atomic/molecular properties.

For example, the speed of a gas molecule is a molecular property. The macroscopic manifestation of the molecular speed is the temperature of the gas. Thus temperature, which we can observe on the macroscopic scale and can in fact discuss without any reference to molecules, is a reflection of average behaviour on the molecular scale. Similarly, the momentum of a gas molecule is a molecular property. The cumulative result of huge numbers of molecules transferring momentum to the container walls during collisions is the macroscopic property called pressure. Again, we may detect and discuss pressure for large samples of gas without any reference to molecules, even though it is a direct consequence of their existence.

As you progress through this text, you will find yourself becoming conversant with both the macroscopic and atomic/molecular viewpoints. It is important that you realise, though, that the molecular interpretation is very often inferred from our observations of macroscopic behaviour. Only recently have we become technologically capable of observing the atomic realm. In doing so, we have found that our inferences about it, made from many observations over many many years, are largely correct. We may define chemistry as the branch of science concerned with the molecular structure of matter, the relation between this structure and the behaviour of matter, and the practical applications of this behaviour.

This definition takes us into the realm of atoms and molecules. As you progress through your study of chemistry, you will become more comfortable in this realm. Throughout the journey, we attempt to keep you focussed on the questions, "What do the molecules look like, what are they doing, and how is this related to the properties and behaviour of the material in which we are interested?" By the end, you should have a good sense of the manner in which chemists look at the world.

THE EVOLUTION OF THE SCIENCE OF CHEMISTRY

It is assumed from the outset that you know a bit about matter, so we will spend minimal time on concepts that you have probably learned before. We will devote just a few lines to the concept of a pure substance, which may be either an element or a compound.

A sample of matter is considered to be a pure substance if it cannot be separated into two or more distinctly different components by physical methods. A physical method can be something as simple as using a pair of forceps (tweezers) to separate two clearly different types of particles from a sample, to something as sophisticated as instrumental chromatography. A pure substance has exactly the same composition throughout its bulk; this composition remains unchanged no matter what physical operations we carry out on the substance.

An element is a fundamental pure substance. It cannot be separated into other pure substances by chemical means (such as decomposing by heat or light). In modern terms, it consists of just one kind of atom. A compound is a combination of two or more elements with a specific, reproducible composition, which is the same throughout the bulk of any sample of the compound. In modern terms, a compound consists of combinations of atoms of two or more elements in very specific arrangements.

It is only rarely possible to do meaningful experiments on matter that is not pure. Consequently, the concern of the chemist is with pure substances. The chemist is a manipulator of matter; s/he must start from a known point, with a known and pure material, to manipulate successfully.

Modern science began when people began to do experiments. Experiments consist of making observations of various types on well-controlled, well-defined situations and processes. From careful experiments it is possible to notice regularities in nature that are not evident in the erratic and chaotic course of ordinary life. Having done experiments, one can proclaim the regularities with confidence, and attempt to find explanations for them.

One of the first individuals to take the experimental approach was Antoine Lavoisier, a native of France, who at the end of the 18th century carried out a series of well-controlled experiments involving the burning of substances. Lavoisier found that, to within the fairly large margin of error that his equipment required, the total mass of the products of a chemical reaction is the same as the total mass of the reactants; mass is conserved in chemical processes.

Law of Conservation of Mass

In any chemical reaction, the total mass of products is the same as the total mass of reactants.

Thus when a steel garden tool rusts as the iron of which it is composed reacts with oxygen from the air, the mass of rust obtained is equal to the sum of the masses of iron and oxygen used to form it.

Only a few years after Lavoisier did his experiments, Joseph Proust discovered via careful experiments of his own that simple chemical compounds such as sodium chloride (modern name) and water always contain the same elements in the same relative amounts, no matter what the source of the particular sample analyzed.

Thus samples of water from a lake, from a distant stream, and even from the ocean (once impurities are removed) all contain only hydrogen and oxygen, and always contain 8 times more oxygen than hydrogen, by mass. The compound ammonia is formed by reaction of the elements nitrogen (N) and hydrogen (H).

No matter what its source or how it is made, ammonia always contains 82.2 per cent nitrogen and 17.8 per cent hydrogen by mass. The relative amounts of N and H in an ammonia sample are always the same. Observations like these led Proust to propose the

Law of Definite Proportions

The relative masses of the elements in a compound is always the same, regardless of the source of the sample of compound tested. Finally, an interesting corollary to the Law of Definite Proportions became known at about this same time. This corollary, called the Law of Multiple Proportions, is stated (rather clumsily) as follows.

Law of Multiple Proportions:

If two elements, A and B, combine to form two different compounds, the masses of B that combine with a particular mass of A give a whole number ratio. For example, in addition to ammonia, nitrogen and hydrogen form a compound called diazene that contains 93.3 per cent nitrogen and 6.7 per cent hydrogen by mass. The mass of hydrogen that would combine with 1.00g of nitrogen to form diazene is therefore

$$1.00 \text{ g nitrogen} \times (6.7 \text{ g hydrogen}/93.3 \text{ g nitrogen}) = 0.0718 \text{ g hydrogen}$$

Similarly, the mass of hydrogen combining with 1.00 g of nitrogen to form ammonia is

$$1.00 \text{ g nitrogen} \times (17.8 \text{ g hydrogen}/82.2 \text{ g nitrogen}) = 0.217 \text{ g hydrogen}$$

The ratio of these two masses of hydrogen is 0.217/0.0718 = 3, in agreement with the Law of Multiple Proportions. These three laws are the foundation of modern chemistry. They provided the basis for the first credible theory of atoms, proposed by John Dalton.

THE ATOMIC THEORY

In 1803, John Dalton proposed that all matter is composed of atoms. Today we know this to be true, because we have seen atoms using advanced microscopic techniques. The postulates that Dalton proposed, although in some cases not altogether true, remain valid for the most part today. They are given below, with the untrue parts enclosed in square brackets:

- Each element is composed of extremely small, indivisible particles called atoms.
- All atoms of an element are identical.
- Atoms of different elements have different properties (including different mass).
- Atoms of an element are not changed into different

types of atoms by chemical reactions; atoms are neither created nor destroyed in such reactions.

- Compounds are formed when atoms of more than one element combine.
- In a particular compound, the relative number and kind of atoms are constant.

The atomic theory is the single most important idea in science. Its essential statement merits repeating, with due emphasis: All matter is composed of atoms, either single atoms or well-defined combinations of atoms. Dalton's theory explains the Laws of Conservation of Mass and Definite Proportions, both well established before Dalton proposed his atom theory. T

he 4th postulate above recognizes the validity of the law of mass conservation. The Law of definite proportions is recognized in postulates 2, 5, and 6. With Dalton's theory in hand, the interpretation of the Law of Multiple Proportions is clear: there are 3 times as many hydrogen atoms per nitrogen atom in ammonia as there are in diazene. The integral mass ratios strongly imply the existence of atoms with characteristic masses. Since the late 1800's, we have learned that atoms are not indivisible. They are composed of still smaller particles, arranged in an interesting way.

THE NUCLEAR ATOM

Atoms are composed of three smaller particles, called subatomic (or elementary) particles: the electron (e), the proton (p), and the neutron (n). These have the properties given in Table.

Table. Subatomic Particles

Particle	*Mass in grams*	*Mass in amu*	*Charge in electronic charge units*
Electron	9.11×10^{-28}	0.0055	–1
Proton	1.673×10^{-24}	1.0073	+ 1
Neutron	1.675×10^{-24}	1.0087	0

The electron and the proton carry electrical charge. The electron is negatively charged, the proton positively charged, with the magnitudes of the charges being exactly the same at

1.602×10^{-19} coulombs (C). This amount of charge is, for convenience, defined as *1 electronic charge unit*. When we speak of the charge of an electron, we will call it -1, but understand that this actually means -1.602×10^{-19} coulombs.

An atom is electrically neutral; it contains the same number of protons and electrons. The protons and neutrons are concentrated together in a small region of space called the nucleus.

The diameter of the nucleus is about 1×10^{-15} m. These two types of particles are bound together in the nucleus by the so-called nuclear strong force, which is much more powerful than the repulsive electrical force that the protons exert on each other (particles of like electrical charge repel each other).

The nuclear strong force is still incompletely understood, primarily because we are not yet able to create energies in the laboratory sufficient to overcome it.

The electrons are found in a relatively large volume (diameter about 1×10^{-10} m) surrounding the nucleus, and are held to it by the attractive electrical force between positive protons and negative electrons.

The magnitude of this attractive force is given by Coulomb's Law, which states that the force between two electrical charges, q_1 and q_2, is directly proportional to the product of the charges and inversely proportional to the square of the distance between them:

$$F = kq_1q_2/r^2$$

The electrons are often said to "orbit" the nucleus, but this planetary view of the situation is not correct, because it implies that the electrons are behaving like well-defined little balls of matter.

When they are confined in atoms, electrons actually behave more like waves than like particles. They are distributed about the nucleus like standing waves on a string with both ends tied down. It may therefore be best to think of the atom as a tiny massive nucleus surrounded by a somewhat diffuse cloud of electron density. The size of the nucleus is exaggerated in this picture so that you can see it. Drawn to scale, it would be too small to see. Chemistry is concerned with the behaviour of the electrons in the atom, and seldom ventures into the nuclear regime.

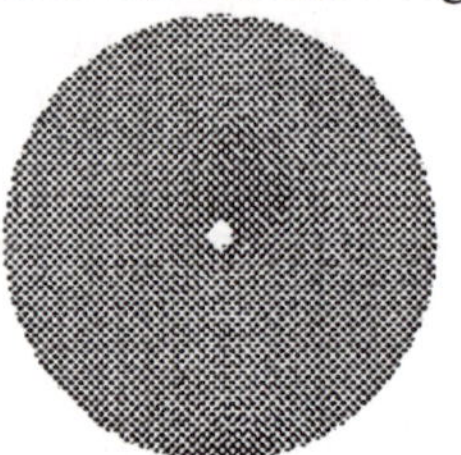

Fig. The Nuclear Atom

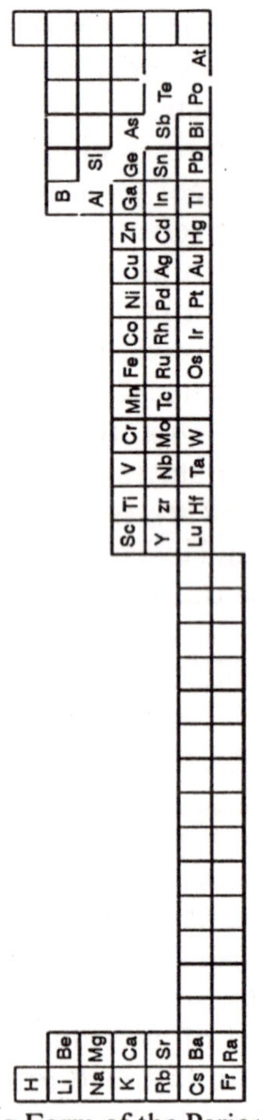

Fig. Long Form of the Periodic Table

However, the electrical pull of the nucleus on the electrons, particularly the outer ones, is of tremendous importance in chemical behaviour. Be mindful of the presence of the nucleus at the centre.

Atoms of different elements are distinguished by the number of protons in the nucleus. Hydrogen atoms have 1 proton in the nucleus, carbon atoms 6, iron atoms 26, etc. The number of protons in the nucleus of the atom of an element is called the atomic number (symbol Z) of the element.

The total number of protons and neutrons in the nucleus is the mass number, A. Although all atoms of a particular element have the same number of protons, they may have different numbers of neutrons. Atoms of an element with different numbers of neutrons are called isotopes. Carbon has 3 isotopes, each with 6 protons and with 6, 7, and 8 neutrons respectively. These have mass numbers 12, 13, and 14 and are distinguished as follows:

$$^{12}_{6}C$$
$$^{13}_{6}C$$
$$^{14}_{6}C$$

An arrangement of the known elements in increasing order of atomic number Z is called the Periodic Table of the Elements. Figure shows the so-called long form of the Periodic Table. The rows of the table are called periods; the columns, numbered 1, 2, 3f through 16f, and 3 through 18, are called groups or families, because all elements in a column exhibit similar chemist3ry.

(For example, the elements in group 1 (except hydrogen) are all metals; react with water to produce hydrogen gas; react vigorously with chlorine to produce salts; and have relatively low melting points.)

The periodic (i.e., repeating) behaviour of the properties of the elements is readily understandable in terms of the quantum mechanical theory of the atom.

ATOMIC MASSES

The concept of atomic mass is crucial to Dalton's atomic theory. With the advent of this theory, chemists became concerned with the masses of atoms and how to measure them. Direct measurement of the masses of single atoms was recognized to be impossible. However, chemists realized that determination of the *relative* masses of the atoms of different elements would allow great progress to be made in understanding the atomic compositions of molecules, even if the absolute masses could not be determined. For example, suppose one knew that a carbon atom was 12 times heavier than a hydrogen atom. Then the fact that a compound contains 12 g of carbon for each 4 g of hydrogen allows the conclusion that the compound contains 4 atoms of hydrogen per atom of carbon.

When Dalton made his proposals, there was no compound in which the numbers of atoms of each type were known. However, combining masses of the elements were known for a number of compounds. Thus water was known to consist of 89 per cent O and 11 per cent H, indicating that in forming water, 8g of O combine with each 1g of H. Although much combining mass data was available, atomic masses could not be assigned to the elements until at least one formula was known. This presented a thorny problem, that was eventually solved by the efforts of Gay Lussac, who worked on the combining volumes of gases; Cannizzaro, and Avogadro.

The Modern Atomic Mass Scale/Mass Spectrometry

We have available to us now a set of atomic masses that are crucial to all quantitative work in chemistry. The masses are based on the arbitrary assignment of a mass of exactly 12.000... atomic mass units to an atom of the $^{12}_{6}C$ isotope. For convenience, this definition gives each element an atomic mass that is numerically close to (but not exactly equal to) its mass number, A (the number of protons and neutrons in the nucleus). The atomic mass unit, symbolized m, is then 1/12 the mass of an atom of ^{12}C. Today we know that 1 m = 1.660 X 10^{-24} g. The units of atomic mass are m/atom. Atomic masses are included with the symbols of the elements on the inside front cover of the text.

A sample of the substance whose mass is to be determined

is vaporized in the ion source, and the gaseous atoms or molecules are bombarded with a beam of high energy particles. These cause the ejection of an electron from several of the atoms/molecules, which become positively charged. The positive ions are accelerated through an electric field of known strength and are then passed through a slit, which focuses them to a narrow beam.

The very precise atomic masses that you find in the periodic table, or in listings of atomic masses, were for the most part not determined from combining mass data. Instead, they were determined by Mass Spectrometry. The operation of a mass spectrometer is illustrated schematically in Figure.

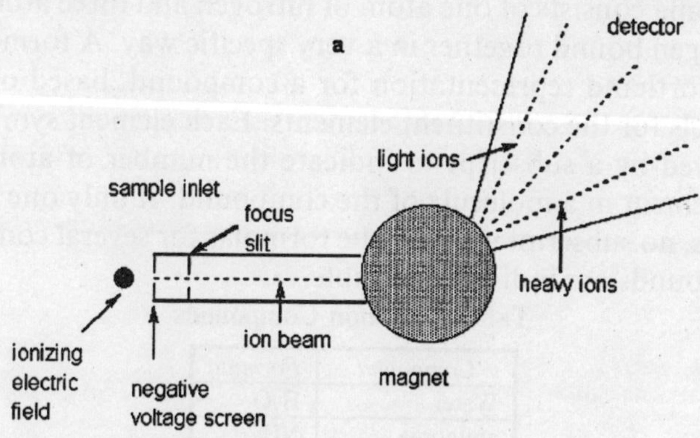

Fig. The Mass Spectrometer

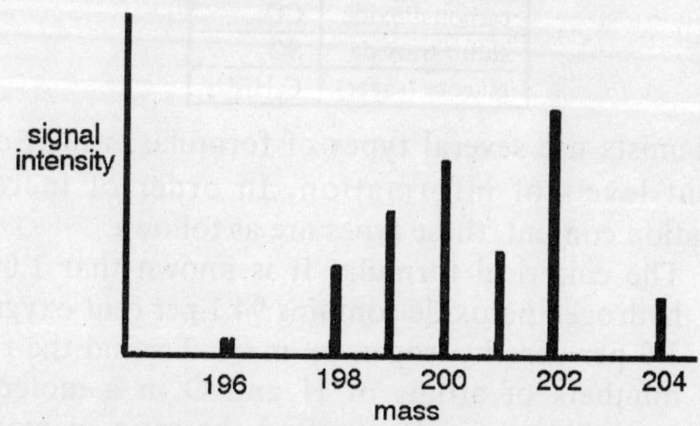

Fig. The Mass Spectrum of Mercury

The beam of positive ions is passed through a magnetic

field of adjustable strength, which causes the beam to follow a curved path. The extent of curvature of the path depends on the mass of the ion, which can be determined from the position at which the beam strikes the detector. A modern mass spectrometer can distinguish between ions that differ in mass by as little as 0.0005 m. The mass spectrum of mercury, Hg, is shown in Figure. Seven isotopes are evident.

FORMULAS

Many chemical compounds consist of molecules. We may define a molecule as a grouping of atoms bound tightly enough together to act as a single entity. For example, a molecule of ammonia consists of one atom of nitrogen and three atoms of hydrogen bound together in a very specific way. A formula is the shorthand representation for a compound, based on the symbols for the constituent elements. Each element symbol is followed by a subscript to indicate the number of atoms of the element in a molecule of the compound. If only one atom occurs, no subscript is used. The formulas for several common compounds are indicated in table.

Table. Common Compounds

Compound	*Formula*
Water	H_2O
ammonia	NH_3
Methane	CH_4
carbon dioxide	CO_2
sulfur trioxide	SO_3
glucose (sugar)	$C_6H_{12}O_6$

Chemists use several types of formulas, which convey different levels of information. In order of increasing information content, these types are as follows.

- The empirical formula. It is known that 1.00 g of hydrogen peroxide contains 94.1 per cent oxygen and 5.9 per cent hydrogen by mass. To find the relative numbers of atoms of H and O in a molecule of hydrogen peroxide, we find the ratio of masses of hydrogen and oxygen in the compound and compare this with the ratio of masses of H and O atoms: mass

H in hydrogen peroxide/mass O in hydrogen peroxide = (# atoms H x atomic mass H)/(# atoms O x atomic mass O) Solving this for the ratio of numbers of atoms, we obtain 1.00. For every atom of O in a molecule of hydrogen peroxide, there is one atom of H. The simplest formula that conveys this information is HO, which is called the empirical formula for hydrogen peroxide.

- Molecular formula. The molecular formula for a compound shows the actual numbers of atoms of each element present in a molecule. The molecular formula for hydrogen peroxide is H_2O_2. The formulas given in Table are all molecular formulas.
- Structural formula. The structural formula for a molecule indicates not only how many of each type of atom is present, but also how the atoms are bonded to each other in the molecule. For example, the structural formulas for hydrogen peroxide and ammonia are shown here.

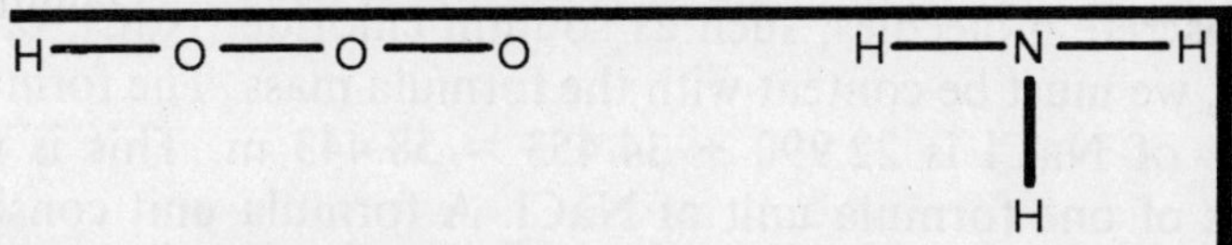

The lines between atoms represent the chemical bonds holding atoms together. The structural formula gives more information than the molecular formula, which in turn is more informative than the empirical formula. The primary goal of a chemist who synthesizes a new compound is to determine the structural formula.

Chemists follow rules when writing formulas for compounds. Unfortunately, different sets of rules are used for compounds of different types. For now, we state two rules that are particularly useful: *For carbon-based compounds, the elements are written in the order, C, H, N, O, others.*

For binary (2-element) compounds, the more positive element is written first. In short order, we will see that in many binary compounds, the more positive element is easily discerned.

Formula Mass and Molecular Mass

Given the formula for a substance, we may define a quantity called the *formula mass* as the sum of the atomic masses of the atoms in the formula. This places a formula unit of a compound on the same mass scale as the atoms of the elements.

Example: Calculate the formula mass of magnesium chloride, $MgCl_2$.

Solution. Add the masses of the atoms in the formula:

FW = atomic mass(Mg) + 2×atomic mass(Cl) = 24.305 + 2(35.453) = 95.211 m

For compounds such as those in Table, consisting of discrete molecules (i.e., individual units that contain a characteristic number of atoms of each constituent element), the formula mass is called the *molecular mass*. The molecular mass is the actual mass of a molecule of the compound, in m.

Thus the molecular mass of a molecule of ammonia, NH_3, is the sum of the masses of 1 N atom and 3 H atoms: 17.031 atomic mass units. For compounds that do not consist of discrete molecules, such as sodium chloride (NaCl, table salt), we must be content with the formula mass. The formula mass of NaCl is 22.990 + 34.453 = 58.443 m. This is the mass of one formula unit of NaCl. A formula unit consists of one Na atom and one Cl atom, as written in the formula.

The Quantitative Implications of Formulas

A chemical formula carries a lot of quantitative information. This is readily illustrated by example.

Example: The molecular formula for acetaminophen, the active ingredient in the pain reliever, tylenol, is $C_8H_9NO_2$. What quantitative relationships are implied by this formula?

Solution. There are a great many such relationships:

- 1 molecule of acetaminophen contains 8 atoms of carbon.
- 1 molecule of acetaminophen contains 9 atoms of hydrogen.
- 1 molecule of acetaminophen contains 1 atom of nitrogen.

- 1 molecule of acetaminophen contains 2 atoms of oxygen.
- For each atom of oxygen in a molecule of acetaminophen, there are 4 atoms of carbon.
- For each atom of nitrogen in a molecule of acetaminophen, there are 9 atoms of hydrogen, 2 atoms of oxygen, and 8 atoms of carbon.

(Note that we would NOT make the statement that, for example, there are 4.5 atoms of H per atom of O because it is meaningless to speak of half an atom. Instead, we say 9 atoms of H per 2 atoms of O.)

- 1 molecule of acetaminophen has a mass of 151.165 m.
- 1 molecule of acetaminophen contains a mass of C of 96.088 m
- 1 molecule of acetaminophen contains a mass of H of 9.072 m
- 1 molecule of acetaminophen contains a mass of N of 14.007 m
- 1 molecule of acetaminophen contains a mass of O of 31.998 m
- For each gram of carbon in acetaminophen, there are 31.998/96.088 grams of oxygen.
- For each gram of nitrogen in acetaminophen, there are 9.072/14.007 grams of hydrogen.

There are many others. Reasoning similar to that in the last two statements was used in discussing the Law of Multiple Proportions earlier.

There are some other important quantitative relationships implied in formulas that we will return to shortly.

IONS

Chemical reactions involve the electrons in the atom, and leave the nuclei untouched. This is because it is relatively easy to either remove electrons from or add them to an atom. Doing so destroys the charge balance of the atom, and produces an electrically charged particle called an ion. Adding one or more electrons to an atom gives a negative entity called a negative ion, or anion.

Removing one or more electrons gives a positive entity called a positive ion, or cation. In both cases, the algebraic difference between the numbers of protons and electrons in the ion is the electrical charge of the ion, in electronic charge units. This is indicated as a right superscript.

Example: Describe the formation of the Na^+, Cl^-, Ca^{2+}, and S^{2-} ions from the atoms.

Solution: Descriptions are in the table.

Atom	*Ion*	*p – e*	*Charge*
Na atom	Na^+ by removal of 1 e	11 – 10 =	1
Cl atom	Cl – by addition of 1 e	17 – 18 =	–1
Ca atom	Ca^{2+} by removal of 2 e	20 – 18 =	2
S atom	S^{2-} by addition of 2 e	16 – 18 =	–2

Having introduced ions, we ask an obvious question: Why does sodium, Na, prefer to lose an electron to give Na^+, while chlorine, Cl, prefers to gain one to form Cl-? The answer is rooted in the detailed structure of the atom. Nonetheless, we can still predict easily which elements tend to form cations and which anions, and further, how many electrons a particular atom is likely to gain or lose, based on the position of the atom in the Periodic Table.

At the right side of the table is a zig-zag line bordering boron on the left and ending between astatine (At) and polonium (Po). This is the boundary between metals (to the left of and below the line) and non-metals (above and to the right). Metals have the following properties:

- A shiny, reflective appearance;
- Ability to conduct heat and electricity;
- Malleability (capable of being hammered into sheets) and ductility (drawn into wires);
- High melting and boiling points (with many exceptions, for example, mercury!)

Non-metals are very different:

- poor conductivity of heat and electricity;
- brittleness — cannot be drawn or hammered without fracturing or crumbling.

- Many are gases at room temperature, and those that are solids are dull in appearance.

Elements adjacent to the boundary line are called semimetals or metalloids, because they have some properties of both under certain conditions.

Based on this classification, we now state some rules about ion formation.

- Metals usually form cations, non-metals anions.
- Metals in groups 1,2, and 13 form cations having 1, 2, and 3 + charges, respectively. Thus the expected charge is given by the last digit of the group number. The following are examples:
- Non-metals tend to form anions with the same number of electrons as the element in group 18 that follows them in the periodic table. The charge on the anion is obtained by subtracting 8 from the last digit of the group number.
- Hydrogen forms the positive ion, H^+. This ion plays a unique and important role in chemistry. Note that, since the hydrogen atom has only one electron, the hydrogen ion is a bare nucleus consisting of a single proton! We will frequently refer to the H^+ ion as the proton.

Element	Group	Predicted Cation
Li	1	Li^+
Be	2	Be^{2+}
Al	13	Al^{3+}
Ba	2	Ba^{2+}
Cs	1	Cs+
Tl	13	Tl^{3+} (also Tl^+)

Element	*Group*	*Group # -8*	*Predicted anion*
N	15	$5-8=-3$	N^{3-}
O	16	$6-8=-2$	O^{2-}
Cl	17	$7-8=-1$	Cl^-
Se	16	$6-8=-2$	Se^{2-}

These rules apply to simple ions, called monatomic ions, formed by adding electrons to or removing them from a single atom. It is also possible to have complex, or polyatomic, ions containing a group of 2 or more atoms, the group as a whole carrying a + or–charge.

POLYATOMIC IONS

A polyatomic ion is a group of atoms bound together as in a molecule, which carries a net + or–charge due to a deficiency or surplus of electrons.

Example. Aluminum sulfate has 2 aluminum and 3 sulfate ions per formula unit. What is its formula?

Table. Common Polyatomic Ions

Name	*Formula*
ammonium	NH_4^+
acetate	$C_2H_3O_2^-$
azide	N^{3-}
carbonate	CO_3^{2-}
hydrogen carbonate	HCO_3^-
hypochlorite	ClO-
chlorite	ClO_2^-
chlorate	ClO_3^-
perchlorate	ClO_4^-
chromate	CrO_4^{2-}
dichromate	$Cr_2O_7^{2-}$
cyanate	OCN-
cyanide	CN-
hydroxide	OH-
nitrite	NO_2^-
nitrate	NO_3^-
permanganate	MnO_4^-
peroxide	O_2^{2-}
phosphate	PO_4^{3-}
hydrogen phosphate	HPO_4^{2-}
dihydrogen phosphate	$H_2PO_4^-$
phosphite	PO_3^{2-}
sulfite	SO_3^{2-}
hydrogen sulfite	HSO_3^-
sulfate	SO_4^{2-}
hydrogen sulfate	HSO_4^-
thiosulfate	$S_2O_3^{2-}$

Solution. Remembering to enclose the polyatomic ion in parentheses, we obtain $Al_2(SO_4)_3$. Aluminum is written first in the formula because it is positive; the sulfate anion is negative.

The ammonium ion, NH_4^+, is a polyatomic cation. The sulfate ion, SO_4^{2-}, is a polyatomic anion. We cannot now state a simple rule for predicting formulas and charges for polyatomic ions. This requires more extensive knowledge of the electronic structure of atoms and the principles of chemical bonding. However, it is important that you be familiar with the formulas and charges of at least the common polyatomic ions. These are listed in Table, which should be committed to memory.

Compounds containing polyatomic ions are common and numerous. Simple examples are sulfuric acid, H_2SO_4; phosphoric acid, H_3PO_4; ammonium perchlorate, NH_4ClO_4; magnesium nitrate, $Mg(NO_3)_2$. The last example shows that when the formula for a polyatomic ion occurs more than once in the formula for a compound, the ion formula is enclosed by parentheses and an appropriate subscript is written after it.

TYPES OF COMPOUNDS

Many compounds consist of discrete molecules, in which atoms are tightly bonded together, but which act as independent units. Examples are water (H_2O), methane (CH_4), ethanol (C_2H_6O), and ammonia (NH_3).

However, many compounds are not molecular. An example is sodium chloride, NaCl, which consists of Na^+ and Cl- ions arranged in a regular, essentially infinite, 3-dimensional array called a crystal lattice.

Each Na^+ ion is surrounded by 6 Cl-, spaced at equal distances from it; similarly, each Cl- is surrounded by 6 Na^+. It is not meaningful to associate any one Na^+ with any one Cl-, so no discrete molecules exist.

It is more proper to consider the entire NaCl crystal as a single giant molecule containing equal huge numbers of Na^+ and Cl- ions.

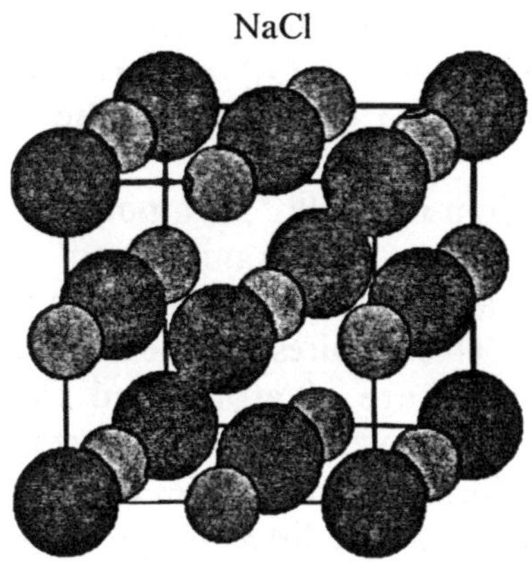

Fig. The Sodium Chloride Lattice

NaCl and other compounds with lattice structures consisting of ions are called ionic compounds. The formulas for such compounds are empirical (not molecular), showing the smallest relative numbers of ions necessary to give electrical neutrality. For these compounds, we calculate a formula mass rather than a molecular mass, and we deal in formula units rather than molecules.

Example. Predict the formula for the ionic compound formed between aluminum and oxygen.

Solution. First predict ion charges: Al^{3+} and O^{2-}, based on periodic table position. Now write an electrically-neutral formula. This requires 2 Al^{3+} for each 3 O^{2-}. Thus

$$Al_2O_3$$

Aluminum is written first in the formula because it is the positive species; oxygen is negative.

Ionic compounds such as NaCl, KBr, and Al_2O_3 are formed by transfer of electrons from one atom to another to form ions. The resulting ions exert strong electrical attractive forces on one another called ionic bonds, which hold them together in the orderly lattice. In contrast, compounds such as water, consisting of discrete molecules, are called molecular (covalent) compounds. Bonds between atoms in these molecules result not from electron transfer, but from electron sharing. The bonds are called covalent bonds.

Given 2 elements, will they form an ionic or a covalent compound? There are 3 simple guidelines that allow such predictions to be made.

- If one element is a metal, the other a non-metal, the resulting compound is ionic. Formulas of the compounds can be predicted as in Example above.
  Examples:

Substance	*Metal*	*Non-metal*
NaCl	Na	Cl
CaF_2	Ca	F
CsI	Cs	I
Al_2S_3	Al	S

- Compounds between metals and complex anions are ionic (however, bonding within the anion is covalent).
- If both elements are non-metals, the compound is covalent.

We have seen that formulas for ionic compounds are easy to obtain. Those for covalent compounds are as well, with an understanding of valence. The valence of an atom is the number of bonds that the atom can form to other atoms. The primary valence for an atom is predictable, just as is ionic charge. We can see this by looking at the formulas for the hydrogen and fluorine compounds of elements in periods 2 and 3 of the periodic table:

Group:	*1*	*2*	*13*	*14*	*15*	*16*	*17*
Period 2 cpd with H:	LiH	BeH_2	BH_3	CH_4	NH_3	OH_2	FH
with F:	LiF	BeF_2	BF_3	CF_4	NF_3	OF_2	FF

The number of bonds formed to hydrogen and fluorine varies in a systematic way from left to right across a row of the periodic table, from 1 at the left, to a maximum of 4 in the middle, and back to 1 at the right. These observations suggest two rules of valence.

- For atoms in groups 1, 2, 13, and 14, the valence is the last digit of the group number. Thus

Element	*Group*	*Valence*
H	1	1
Mg	2	2
Ga	13	3
C	14	4

- For atoms in groups 15 to 18, valence is equal to (8– last digit of group #).

Element	*Group*	*Valence*	
O	16	8-6 = 2	
Cl	17	8-7 = 1	
N	15	8-5 = 3	

Note that the way valence is defined requires it to be a positive number. The rules are stated so as to guarantee this.

With these rules, the formula for water is sensible. Oxygen (valence 2) makes 2 attachments; hydrogen (valence 1) makes 1 attachment. One oxygen binds with 2 hydrogens, as follows:

H–O–H

The formula HO satisfies the valence of hydrogen, but not that of oxygen. It is therefore incorrect. Similarly the formula for CO_2 is sensible:

O=C=O

Carbon has 4 attachments to satisfy a valence of 4; each oxygen has 2 attachments to satisfy a valence of 2.

THE MOLE

The mole is the unit of amount in chemistry. It provides a bridge between the atom and the macroscopic amounts of material that we work with in the laboratory. It allows the chemist to weigh amounts of two substances, say iron and sulfur, such that equal numbers of atoms of iron and sulfur are obtained. A mole of a substance is awkwardly defined as the mass of substance containing the same number of fundamental units as there are atoms *in exactly 12.000...g of* ^{12}C.

Fundamental units may be atoms, molecules, or formula units, depending on the substance concerned. At present, our best estimate of the number of atoms in 12.000...g of ^{12}C is 6.022×10^{23}, a huge number of atoms. This is obviously a very important quantity. For historical reasons, it is called Avogadro's Number, and is given the symbol N_0.

Unfortunately, the clumsy definition of the mole obscures its utility. It is nearly analogous to defining a dozen as the mass of a substance that contains the same number of fundamental units as are contained in 733 g of Grade A large eggs. This definition completely obscures the utility of the dozen: that it is 12 things! Similarly, a mole is N_0 things.

The mole is the same kind of unit as the dozen — a certain number of things. But it differs from the dozen in a couple of ways. First, the number of things in a mole is so huge that we cannot identify with it in the way that we can identify with 12 things.

Second, 12 is an important number in the English system of weights and measures, so the definition of a dozen as 12 things makes sense. However, the choice of the unusual number, 6.022×10^{23}, as the number of things in a mole seems odd. Why is this number chosen? Would it not make more sense to define a mole as 1.0×10^{23} things, a nice (albeit large) integer that everyone can easily remember? To understand why the particular number, 6.022×10^{23} is used, it is necessary to resurrect an older, in some ways more sensible and useful, definition of the mole, which is grounded in the atomic mass scale addressed above.

The atomic mass scale defines the masses of atoms relative to the mass of an atom of ^{12}C, which is assigned a mass of exactly 12.000.. atomic mass units (m). The number 12 is chosen so that the least massive atom, hydrogen, has a mass of about 1 (actually 1.008) on the scale. The atomic mass unit is a very tiny unit of mass appropriate to the scale of single atoms. Originally, of course, chemists had no idea of its value in laboratory-sized units like the gram. The early versions of the atomic mass scale were established by scientists who had no knowledge of the electron, proton, or neutron. When these were discovered in the late 19th and early 20th centuries, it turned out that the mass of an atom on the atomic mass scale was very nearly the same as the total number of protons and neutrons in its nucleus. This is a very useful correpondence, but it was discovered only after the mass scale had been in use for a long time.

In their desire to be able to count atoms by weighing, chemists gradually developed the concept of the "gram-atomic mass", which was defined in exact correspondence with the atomic mass scale:

1 atom of ^{12}C has a mass of 12.000 amu

1 gram-atomic mass of ^{12}C has a mass of 12.000 g

Thus the gram-atomic mass of an element was defined as the atomic mass of the element, expressed in grams. Because the atomic mass scale is numerically preserved in the definition of gram atomic mass, 1 gram-atomic mass of any element could be immediately determined as the atomic mass in grams.

Thus 1 gram-atomic mass of sulfur is 32.06 g; 1 gram-atomic mass of hydrogen is 1.008 g, and so on. Analogous terms, such as gram-molecular mass for the molecular mass of a compound expressed in grams, were similarly used. However, having to use a different term depending on whether elements or compounds were being discussed was awkward and inconvenient. For this reason, the term "molar mass (abbrev MM)" (the mass of 1 mole) was adopted to signify the atomic, molecular, or formula mass of a pure substance expressed in grams.

Alternative definition of the mole: a mole of a pure substance is the atomic, molecular, or formula mass of the substance, expressed in grams.

Thus one mole of ethyl alcohol, C_2H_6O, is 46.069g. One mole of water is 18.015 g. If we mix 46.069 g of ethyl alcohol with 18.015 g of water, we can be assured that the mixture contains 1 molecule of ethyl alcohol per molecule of water. Further, we will know that there are 2 atoms of C and 8 atoms of H per each 2 atoms of O. Thus the mole allows us to measure convenient amounts of material containing known numbers of atoms; i.e., it allows us to count atoms.

The mole enables us to count atoms in the laboratory: The mole is useful whether or not we know how many atoms of carbon-12 there are in 12.000 g of carbon-12. If we measure one mole of iron and one mole of sulfur, we know that these two samples contain the same number of atoms. This is the important aspect of the mole.

How many atoms there are in a mole is of subsidiary

importance. Nonetheless, it has become possible to determine this number. It is, of course, 6.022×10^{23} atoms per mole. We thus see that this number is simply a *consequence* of the choice that 1 mole be the formula mass in grams. It is very nice that we know it; but we do not need to know it for the mole to be useful.

I would even go so far as to say that the modern definition of the mole in terms of a certain number of atoms of ^{12}C is unfortunate, in that it suggests that the number, 6.022×10^{23} things/mole, must be used in any and every calculation involving moles!

In practice, we seldom need to know how many atoms or molecules we are working with, so in mole calculations the number 6.022×10^{23} is rarely used. What is invariably used is the fact that 1 mole of substance is its atomic, formula, or molecular mass in grams.

The dual definitions of the mole can be used to find the mass of 1 m expressed in g. Exactly 12 g of carbon contains N_o atoms, each with a mass of exactly 12 m. In equation form,

N_o atoms/mole$\times$ 12 m /atom = 12 g/mole

Simplifying, we obtain N_o m = 1 g. It follows that 1 m = $1/N_o$ g = 1.660×10^{-24} g. Avogadro's number is an experimentally measured quantity. Although we are confident that we know its value quite well, some future experiment may cause us to make a small revision in the number. By necessity, the mass of 1 m in grams will change accordingly.

This is not worrisome, because neither number is crucial to the utility of the mole. The importance of the mole concept can be summed up as follows: any statement that can be made about the number of atoms of an element in a molecule or formula unit of a substance can also be made about the number of moles of an element in a mole of the substance. This is true because 1 mole of substance contains N_o atoms, molecules, or formula units of substance. Based on the formula for glucose, $C_6H_{12}O_6$, we can make the following statements:

- 1 molecule of glucose contains 6 atoms of C, 12 atoms of H, and 6 atoms of O.

- 1 mole of glucose contains 6 moles of C atoms, 12 moles of H atoms, and 6 moles of O atoms.
- 10 molecule of glucose contains 60 atoms of C, 120 atoms of H, and 60 atoms of O.
- 10 moles of glucose contains 60 moles of C, 120 moles of H, and 60 moles of O atoms.
- Any amount of glucose contains equal numbers of C and O atoms, and twice this number of H atoms.
- Any amount of glucose contains equal numbers of moles of C and O atoms, and twice this number of moles of H atoms.
- N_0 molecules of glucose contains $6 \times N_0$ atoms of C, $12 \times N_0$ atoms of H, and $6 \times N_0$ atoms of O.

You might now look back at Example and apply these ideas to the formula for acetaminophen. Some examples will familiarize you with the mole.

Example. How many moles of Fe are in 5.6 g Fe? How many Fe atoms are contained in the sample?

Solution. By definition, 1 mole of Fe is 56.0 g. 5.6 g Fe is therefore 0.1 mole of Fe. The number of Fe atoms in the sample is 0.1 mole $\times$ 6.022 $\times$ 10^{23} atoms/mole = 6.022×10^{22} atoms.

5.6 g of iron is not much iron. However, even this small amount contains a huge number of iron atoms.

Example. How many sulfur atoms are in 1.56 g sulfur?

Solution. We can calculate the number of moles of sulfur from the molar mass and the given mass. Once we have this, the number of atoms is obtained from Avogadro's Number, N_0.

moles S = 1.56 g/(32.06 g/mole) = 0.0487 mole

atoms S = 0.0487 moles S$\times$ 6.022×10^{23} atoms/mole = 2.93×10^{22}

Example. What is the mass in g of 1 atom of sodium?

Solution. If we know the mass of 1 mole of Na, and how many atoms are in a mole, the mass of a single atom should be easy to obtain:

mass Na atom = 22.99 g Na/mole$\times$ 1 mole/6.022×10^{23} atoms = 3.817×10^{-23} g.

Example. What mass of sulfur contains the same number of moles as are in 10.0 g Fe?

Solution. Figure the number of moles of Fe. This is the desired number of moles of S. Convert moles of S to mass of S using the molar mass.

moles Fe = 10.0 g Fe/(55.85 g Fe/mole) = 0.1791 moles
moles S = moles Fe = 0.1791
g S = 0.1791 moles S× 32.06 g S/mole = 5.71 g S

Example. Hemoglobin is the oxygen-carrying protein of most mammals. Each molecule of hemoglobin contains 4 atoms of iron. The molar mass of hemoglobin is about 64000 g/mole. How many moles of iron are contained in 0.50 moles of hemoglobin? Calculate the number of iron atoms in 0.128 g of hemoglobin.

Solution. Based on its molar mass, hemoglobin is clearly a large molecule containing many atoms. We are not told what the atoms are, nor how many of each there are. However, we are told that each molecule of hemoglobin contains 4 atoms of iron. We can write the formula for a molecule of hemoglobin as follows:

$$Fe_4X$$

where X represents the collection of all other atoms present. What we can say about molecules and atoms, we can say about moles. Thus 1 mole of hemoglobin contains 4 moles of iron. Similarly, 0.50 moles of hemoglobin contains 4×0.50 = 2.00 moles of iron.

To obtain the second required answer, we convert mass of hemoglobin to moles hemoglobin using the molar mass: moles Fe_4X = 0.128 g Fe_4X/64000 g/mole = $2.00X10^{-6}$ moles

This contains $4\times2.00\times10^{-6}$ moles of iron. The number of iron atoms is obtained using Avogadro's number:

Number Fe atoms = $4\times2.00\times10^{-6}\times6.023\times10^{23} = 4.82\times10^{18}$ Fe atoms. It is interesting that we do not need to know what the other atoms are in the hemoglobin molecule, much less how many of them there are. Knowledge of the number of iron atoms per molecule of hemoglobin is enough.

For oxygen, which exists in nature as diatomic molecules, O_2, the statement "a mole of oxygen" is ambiguous. Does it mean a mole of oxygen molecules or a mole of oxygen atoms? These are different things. A mole of oxygen molecules contains 2 moles of oxygen atoms. For elements that exist as molecules, it is best to explicitly state whether molecules or atoms are meant. Thus "1 mole of oxygen molecules" means 6.022×10^{23} O_2 molecules, or $2 \times 6.022 \times 10^{23}$ O atoms; "1 mole of oxygen atoms" means 6.022×10^{23} O atoms.

STOICHIOMETRIC CALCULATIONS FOR COMPOUNDS

When a chemist speaks of "stoichiometry", s/he means the quantitative mass relationships that govern the formulas of and the reactions between chemical substances. Stoichiometry rests on the validity of the Law of Conservation of Mass. A fundamental property of a compound is that it contains its constituent elements in definite proportions by mass.

The per cent composition of a compound is a statement of the mass percentages of the elements present in the compound. From the formula for a compound, per cent composition is readily calculated. More importantly, knowledge of per cent composition leads to the formula. The following examples illustrate this two-way street.

Example. What is the per cent composition of nitrogen dioxide, NO_2?

Solution. From the formula, we calculate the mass of one molecule of NO_2. This mass contains the mass of 1 atom of N. The mass per cent N is thus

per centN = $AW(N)/MW(NO_2)\times 100$ = (14.0 m/46.0 m)$\times$100 = 30.4 per cent

per centO = $2\times AW(O)/MW(NO_2)\times 100 = (32.0/46.0)\times 100$ = 69.6 per cent

The sum of the mass percentages of N and O is 100 per cent, as it must be. In general, the percentage by mass of an element in a compound is obtained by multiplying the molar mass of the element by the number of atoms of the element in 1 molecule of compound, dividing by the molar mass of the compound, and multiplying by 100.

Example. The per cent composition of a sample of the mineral, bustamite, is determined by elemental analysis to be

per centCa = 16.21
per centMn = 22.23
per centSi = 22.72
per centO = the remainder
What is the empirical formula of bustamite?

Solution. The approach to this problem is less obvious than in the previous example. The general strategy is as follows. Knowledge of per cent composition allows calculation of the masses of calcium, manganese, silicon, and oxygen in any specified mass of compound. Since this mass is not important, we specify a convenient value — 100 g. Once the mass of each element in this amount of bustamite is known, the masses can be converted to moles by division by the molar masses. The result is the number of moles of each element in the given mass of compound. The relative numbers of moles are then converted to whole numbers by dividing by the smallest number of moles. The flow is as follows:

per centcomp → mass of each element in 100 g cpd → moles each element in 100g cpd → relative numbers of moles in integer form → empirical formula

mass Ca in 100g cpd = 100 g cpd × 0.1621 g Ca/1 g cpd = 16.21 g Ca

mass Mn = 100 g cpd× 0.2223 g Mn/1 g cpd = 22.23 g Mn

mass Si = 22.72 g Si

mass O = 38.84 g O

(The per cent O is determined by subtracting the sum of the other percentages from 100.)

Now convert to moles:

moles Ca = 16.21 g C/(40.08 g Ca/mole) = 0.4044 mole Ca

moles Mn = 22.23 g Mn/(54.938 g Mn/mole) = 0.4046 mole Mn

moles Si = 22.72 g Si/(28.086 g/mole) = 0.8089 mole Si

moles O = 38.84 g O/(15.999 g/mole) = 2.428 mole O

Divide each of these numbers by the smallest of them, 0.4044:

relative moles Ca = 0.4044/0.4044 = 1

relative moles Mn = 0.4046/0.4044 = 1.00
relative moles Si = 0.8089/0.4044 = 2.00
relative moles O = 2.428/0.4044 = 6.00

The empirical formula of bustamite is $CaMnSi_2O_6$.

Note: Calculation of the relative numbers of moles rarely gives exactly integers, because the per cent composition is determined experimentally and is therefore subject to error. However, since we know that atoms must occur in formulas in integer numbers, it is often obvious how to round off the results.

Using the per cent composition to determine the formula is one of the first things that a chemist does with a newly synthesized chemical compound. It is a very important initial step in determining the structural details of the new substance.

CHEMICAL REACTIONS.

A chemical reaction is a process involving changes in atomic arrangements, accompanied by changes in energy. In a chemical reaction, substances are converted into new substances. Some examples of chemical reactions follow:

- Metallic zinc and sulfur (a yellow powder) react when heated to produce zinc sulfide, a white salt.
- Hydrogen gas burns in air to produce water, with the evolution of heat and light. This reaction powers the liftoff of the space shuttle, and may soon power automobiles and heat homes.
- Carbon dioxide and water react in the presence of sunlight and chlorophyll to produce oxygen gas and sugar (the photosynthesis reaction). All life on earth depends on this process.

These processes involve rearrangements of atoms (for example, in c), the carbon atoms that begin in CO_2 end up in glucose molecules!) and all produce or require heat or light energy.

In two of these examples we can tell visually that reaction has occurred. In a), yellow S disappears and is replaced by white ZnS; in b) we can see the light and feel the heat; however, in c), nothing happens visually. We cannot see photosynthesis occurring. Thus although a visible change often accompanies a reaction, it need not.

To describe a reaction more concisely and completely than the word descriptions above, chemists use chemical equations. The chemical equation for reaction a) is

$$Zn + S \rightarrow ZnS$$

Note that 1) chemical formulas are used to represent substances that react and form; 2) Substances that react (the reactants) are written to the left of the arrow; those formed (the products) are written to the right; 3) reactants (and products) are separated by + signs, which conveys that reactants combine or react together; 4) reactants are separated from products by an arrow pointing from the former to the latter.

The arrow is read "yield" or "produce"; 5) the number of atoms of each element in the products is the same as the number of atoms of the element in the reactants (the Law of Conservation of Mass).

The chemical reaction for photosynthesis is written below.

$$CO_2 + H_2O \rightarrow C_6H_{12}O_6 + O_2$$

However, the equation is incomplete, because the numbers of atoms on left and right are not the same; the equation is not balanced. Balancing is accomplished by placing numerical coefficients (called stoichiometric coefficients) in front of the formulas to equalize the numbers of atoms of all types on both sides of the arrow.

The stepwise approach for the photosynthesis reaction is illustrated below.

- As written, the reaction shows 1 C atom on the left, 6 on the right. To equalize the number of C atoms, coefficient 6 is placed in front of CO_2:
  $$6CO_2 + H_2O \rightarrow C_6H_{12}O_6 + O_2$$
- There are 2 H atoms left, 12 right. Balance H by placing coefficient 6 before H_2O:
  $$6CO_2 + 6H_2O \rightarrow C_6H_{12}O_6 + O_2$$
- There are 18 O atoms left, only 8 right. Place coefficient 6 before O_2 on the right:

$$6CO_2 + 6H_2O \rightarrow C_6H_{12}O_6 + 6O_2$$

Balancing is now complete. Note that it is never permissible to balance a chemical equation by altering the subscripts in the chemical formulas, because the subscripts are characteristic of the chemical species involved. Altering them changes.

The identity of the species! Balancing can be done only by altering the stoichiometric coefficients. Our reading of the balanced equation is "6 molecules of CO_2 react with 6 molecules of H_2O to produce 1 molecule of glucose and 6 molecules of O_2." Since one mole of any substance contains N_0 molecules, we may also read it as "6 moles of CO_2 react with 6 moles of H_2O to produce 1 mole of glucose and 6 moles of O_2."

Finally, it is sometimes useful to indicate the physical states (solid, liquid, or gas) of reactants and products in the equation, and/or the reaction conditions. We indicate physical state using symbols in parentheses following the formulas. Common symbols are (g), (l), (s), and (aq) for, respectively, gas, liquid, solid, and aqueous (water) solution. Reaction conditions are usually written either above or below the arrow. Photosynthesis occurs at ambient temperature and pressure, in the presence of sunlight. We indicate this as shown below.

(1-8-1):	$6CO_2(g)$ +	$6H_2O(l)$	T = 298 K, P = 1 atm ------------------> light	$C_6H_{12}O_6(s) + 6O_2(g)$

The Chemical Equation as a Recipe

Frequently, people who are first learning about chemical equations interpret the meaning of the equation incorrectly. In this section, we will attempt to concretize chemical equations, which admittedly relate symbols of unseeable atoms and molecules and are therefore abstract, by drawing an analogy between a chemical equation and a recipe. The analogy is very close; all of the important quantitative aspects of chemical reactions apply equally to recipes. A recipe is a prescription for a process by which specified relative amounts of ingredients are transformed into a desired product. It is a written representation of the process of producing the product. Similarly, a chemical equation is a description of a process by which appropriate relative amounts of reactants are transformed into products. It is a written representation of the process of chemical change. A recipe for a simple fruit salad is given below:

1 apple + 2 oranges + 10 grapes → 1 fruit salad

The recipe specifies the relative numbers of ingredients needed per fruit salad to be made. It tells us the relative numbers of apples, oranges, and grapes to be used. However many fruit salads we plan to make, we must use twice as many oranges as apples, and 10 times as many grapes as apples. The number of fruit salads that result will be the same as the number of apples that we started with.

The equation very definitely does NOT say anything about the numbers of apples, oranges, and grapes that we actually have on hand. Thus we might purchase a bag of 10 apples, a bag of 10 oranges, and a bag of 200 grapes. The ratios of these numbers do not correspond to the ratios of the coefficients in equation.

The coefficients tell us that in making fruit salads, we must use 2 oranges and 10 grapes for each apple that we use. How many fruit salads we can make from our purchased amounts depends on which ingredient we run out of first. For this particular situation, we can make only 5 fruit salads, because we will run out of oranges first. 5 apples and 150 grapes will be left over. If we want more fruit salads, we must buy more oranges.

Chemical equations are exactly like recipes. Thus equation says that for every 6 molecules of CO_2 we use, we must also use 6 molecules of water. We will produce 1 molecule of glucose and 6 molecules of O_2. Note that after reaction, we no longer have the water and the CO_2; we now have glucose and O_2. If we happen to have on hand 20 molecules of CO_2 and 12 molecules of H_2O, then we can produce only two molecules of glucose using our recipe, because we will run out of H_2O. We expect to have 8 molecules of CO_2 left over. The equation describes a conversion process. Chemical equations, like recipes, are usually written according to certain agreed-upon rules (this is probably less true for recipes than for equations) Thus, a recipe for devils food cake specifies the amounts of ingredients needed for 1 cake. Similarly, chemists agree that they will write equations such that all the coefficients are the smallest possible whole numbers consistent with atom balance.

However, just as we can scale the amounts in the recipe to make any number of cakes desired, we can scale the amounts of reactants used in a chemical reaction. Thus in carrying out a chemical process we are not restricted to the specific amounts indicated in the equation (or in cooking, to the specific amounts in the recipe); we can scale the amounts by a factor that is convenient for us. Green plants carry out photosynthesis on a scale of tons; however, the process proceeds according to equation. That is, for each mole of CO_2 used, one mole of H_2O is used, and 1/6 mole of glucose and 1 mole of O_2 are formed.

Most people find this type of reasoning easy to understand in terms of a recipe, because recipes are familiar, and concrete. The reasoning works the same way for chemical reactions; but it is complicated by the abstract aspect of the formulas. If you find yourself bogged down by the abstraction, ground yourself by thinking of the recipe analogy.

Classification of Chemical Reactions

There are literally millions of known chemical reactions. To learn them all individually is a hopeless task. Instead, we attempt to classify reactions into a manageably small number of categories, and learn the fundamental characteristics of each category. We can then make statements about a reaction category that apply to all individual reactions in the category. In this text, we will be concerned with three major reaction types that we will now define and describe.

We are not yet equipped to fully understand certain aspects of these reactions; however, it is best to begin practicing the recognition of reaction type as early as possible, even though it may seem difficult at first. It will become easier. The following three classes of chemical reactions will be discussed in this text:

- Electron Transfer reactions (also known as oxidation-reduction, or redox, reactions)
- Proton Transfer reactions (also known as acid-base reactions)
- Double Displacement reactions

We will define each type, and briefly discuss its characteristics.

Electron Transfer Reactions. These are reactions in which electrons are transferred from one atom, molecule, or ion to another atom, molecule, or ion. The substance that provides the electrons becomes more positive as a result of the process; the substance that receives the electrons becomes more negative. Some electron transfer reactions are very easy to recognize. For example, the reaction of sodium and chlorine in 1-8-3 is an electron transfer process, because sodium becomes more positive and chlorine more negative in the process:

$$2Na(s) + Cl_2(g) \rightarrow 2NaCl(s)$$

Sodium is in the elemental (uncharged) form prior to reaction; after reaction, it exists as Na+ ions in NaCl. Clearly, each sodium atom has lost one electron. At the same time, each Cl atom of Cl_2 has gained an electron to become Cl-. The reaction of magnesium and oxygen is also an electron transfer process:

$$2Mg(s) + O_2(g) \rightarrow 2MgO(s)$$

Each magnesium atom loses two electrons; each oxygen atom accepts two electrons; electrons are therefore transferred from Mg to O.

Many reactions involve electron transfer. We recognize only two subcategories, which are exemplified by reaction the reaction of a metal with a non-metal; and the reaction of an element with oxygen (combustion). We will discuss combustion below.

Proton Transfer Reactions. These are processes in which a proton, H^+, is transferred from one species to another. The species losing the proton is called an acid; the species gaining the proton is called a base. An example of a proton transfer reaction is shown in reaction below.

$$HCl(aq) + Mg(OH)_2(ag) \rightarrow MgCl_2(aq) + HOH\ (H_2O)$$

Here a proton, H^+, is transferred from HCl to the OH- portion of magnesium hydroxide. Water is produced by combination of H^+ and OH-; magnesium chloride is also produced. Similarly, 1-8-6 is a proton transfer process:

$$H_2SO_4(aq) + Na_2CO_3(aq) \rightarrow Na_2SO_4(aq) + H_2CO_3(aq)$$

Here H_2SO_4 transfers two protons to CO_3^{2-} to produce H_2CO_3; sodium sulfate is also produced. If protons are transferred in a reaction, the reaction is an acid-base process.

Double Displacement Reactions. In a double displacement process, the positive and negative parts of two reacting substances exchange: the positive part of one substance ends up with the negative part of the other. For example, the reactions below are double displacement processes:

$$NaCl(aq) + AgNO_3(aq) \rightarrow NaNO_3(aq) + AgCl(s)$$

$$BaCl_2(aq) + K_2SO_4(aq) \rightarrow KCl + BaSO_4(s)$$

You may notice that the proton transfer reactions are also double displacement reactions. This is certainly true. But because acid-base processes play such a central role in chemistry, chemists choose to classify them separately. We will adhere to this convention. Combustion Reactions. In this section, we introduce and discuss an important category of chemical reactions: combustion reactions.

A combustion reaction involves the reaction of a substance, element or compound, with oxygen, usually with the accompanying production of heat or light, or sound energy. The balanced equations representing the combustion reactions of elemental phosphorus and of methane are shown in equation form below:

$$4P(s) + 5\ O_2(g) \rightarrow P_4O_{10}(s)$$

$$CH_4(g) + 2O_2(g) \rightarrow CO_2(g) + 2H_2O(g)$$

P_4O_{10} is called the oxide of phosphorus; similarly, CO_2 is an oxide of carbon, and H_2O is the oxide of hydrogen. Using this terminology, we generalize from these typical combustion processes as follows:

Element + $O_2 \rightarrow$ *oxide of element*

Compound + $O_2 \rightarrow$ *oxides of all elements in the compound*

These generalizations give us tremendous predictive power, as the next example shows.

Example. Predict the products of the following combustion reactions, and balance the resulting equations.

$$S + O_2 \rightarrow$$
$$Fe + O_2 \rightarrow$$
$$H_2 + O_2 \rightarrow$$
$$C_8H_{18} + O_2 \rightarrow$$
$$C_6H_{12}O_6 + O_2 \rightarrow$$
$$NH_3 + O_2 \rightarrow$$

Solution. The first 3 reactions involve combustion of an element; the product is the oxide of the element:

$$S + O_2 \rightarrow SO_2$$

Predicting SO_3 as the product of combustion of sulfur is also reasonable, although it is SO_2 that is actually formed. Further oxidation to SO_3 occurs only very slowly unless a suitable catalyst (a substance that speeds a reaction) is present.

$$4Fe + 3\,O_2 \rightarrow 2Fe_2O_3$$
$$2H_2 + O_2 \rightarrow 2H_2O$$

The remaining three reactions are those of compounds with O_2. We expect to produce the oxides of all of the elements in the compound:

$$2C_8H_{18} + 2O_2 \rightarrow 16CO_2 + 18H_2O$$
$$C_6H_{12}O_6 + 6O_2 \rightarrow 6CO_2 + 6H_2O$$

The oxide of oxygen is O_2. We could place it on the products side, but would have to remove it again in simplifying the equation. The simplest approach is to ignore it.

$$4NH_3 + 5O_2 \rightarrow 4NO + 6H_2O$$

There are a number of oxides of nitrogen, including N_2O, NO, NO_2, N_2O_3, and N_2O_5. NO is the product actually formed. Any of the others would be a reasonable prediction, however.

Each of the reactions is of great industrial, technological, or biological importance. The reaction of sulfur with oxygen is the first step in the industrial process for the manufacture of sulfuric acid. Sulfuric acid is produced in larger amount than any other chemical substance, and is used in most other industrial chemical processes.

The reaction of sulfur with oxygen also occurs when coal is burned to power the generators in electrical power plants. Although coal contains only a small (and variable) amount of sulfur, coal is burned on such a large scale that significant amounts of SO_2 are released into the atmosphere. Conversion of SO_2 to SO_3 is catalyzed by particulate matter suspended in the air; and SO_3 then combines with water to give sulfuric acid:

$$SO_3 + H_2O \rightarrow H_2SO_4$$

Sulfuric acid is the major component of acid rain, which destroys forests, aquatic life in lakes and rivers, and man-made objects such as statues and buildings. The acid rain problem remains to be solved.

The reaction of iron with oxygen is better known as rusting. Although rusting is slow, it leads to the inevitable destruction of iron- and steel-based construction. The rusting of iron and steel is an example of corrosion. Corrosion is responsible for many millions of dollars in damage to the infrastructure of the United States each year.

Reaction of hydrogen with oxygen is used to power the Space Shuttle, the centerpiece of the space programme of the National Aeronautics and Space Administration. Hydrogen is used as the fuel. Unlike other combustion fuels, hydrogen is very desirable because the only product, other than the desired energy output, is water, which of course does no environmental damage.

Currently, although it is possible to produce hydrogen gas from water, it cannot be done economically. When an economical process is developed, hydrogen may be used for less esoteric purposes, such as heating homes and powering automobiles.

The reactant, C_8H_{18}, in Reaction is a major constituent of gasoline, which is used to power the internal combustion engines of automobiles, lawn mowers, snow blowers, jet skis, and the like. Combustion of gasoline releases a large amount of energy, which, when controlled, can be converted effectively to useful work. Gasoline is an example of a hydrocarbon, a compound containing only the elements carbon and hydrogen. There are many known hydrocarbons, the simplest being CH_4 (methane), which is a gas under ordinary conditions. Gasoline, with 8 carbons, is a liquid; and hydrocarbons with even bigger molecules (e.g., $C_{20}H_{42}$) are solids. Methane is used as heating fuel in homes, under the name "natural gas." It is so-called because it is found trapped in huge underground deposits, from which it can be released and piped over long distances.

Glucose, $C_6H_{12}O_6$, and its close relatives are the primary biological fuels. Controlled combustion of glucose provides the energy to power the living cell. Almost all living organisms, plant and animal, carry out some version of equation.

The reverse of, in which carbon dioxide and water are converted to glucose and oxygen, is photosynthesis. This reaction, driven by the energy from sun light, is performed by green plants. Thus these organisms produce their own food via photosynthesis, then consume it.

Finally, the combustion of ammonia, NH_3, is carried out as the first step in the industrial synthesis of nitric acid, HNO_3. Nitric acid is a commercially important substance, and is made in very large quantities each year.

A currently very important chemical problem is to develop inexpensive methods for making NH_3. The currently used process has been in place since about 1915, and is very expensive.

Introduction to Stoichiometry of Chemical Reactions.

Before leaving this topic, we explore some of the quantitative ideas from the previous section. Recall that the important ideas there were

- that equations for chemical reactions provide recipes for converting reactants into products;
- that the coefficients in the equation relate the amounts of reactants that combine to the amounts of products that form.
- that the coefficients indicate both the relative

numbers of molecules, atoms, or formula units AND the relative numbers of moles of reactants and products.

Example. For the photosynthesis reaction, which is the reverse of reaction, make as many quantitative statements as possible about the numbers of molecules and moles of reactants and products.

$$6CO_2(g) + 6H_2O(g) \rightarrow C_6H_{12}O_6(s) + 6O_2(g)$$

Solution. There are so many statements that can be made that we will cover only a small fraction of them. However, let's try:

- In order to convert 6 molecules of CO_2 to glucose, 6 molecules of H_2O are required; when reaction takes place, CO_2 and H_2O disappear and are replaced by 1 molecule of glucose and 6 molecules of $O_2(g)$.
- To convert 6 moles of CO_2 to glucose, 6 moles of H_2O are required; when reaction takes place, 1 mole of glucose and 6 moles of $O_2(g)$ are formed.
- When 1 mole of CO_2 reacts, 1 mole of H_2O must also react. 1/6 mole of glucose and 1 mole of $O_2(g)$ are produced.
- When x moles of CO_2 reacts according to, x moles of H_2O also react; these disappear and are replaced by x/6 moles of glucose and x moles of O_2.
- All of the carbon atoms that start out in CO_2 end up in glucose. Since each CO_2 molecule contains 1 carbon atom and each glucose molecule contains 6 carbon atoms, 6 CO_2 molecules are required to make 1 glucose molecule.
- All of the hydrogen atoms that start out in water end up in glucose. Because each H_2O molecule contains 2 hydrogen atoms and each glucose molecule contains 12, 6 molecules of water are required to supply the hydrogen atoms for 1 glucose molecule.
- If we have on hand 0.25 moles of CO_2 and want to use it to make glucose, we must provide 0.25 moles of water. We will obtain 0.25/6 moles of glucose, and 0.25 moles of water. The total number of atoms of

each type is the same in the reactants and products. This is the Law of Conservation of Mass.

- The number of moles of each type of atom is the same in reactants and products.

This is sufficient to illustrate the types of things we can say. It is important to be comfortable with the information obtainable from a chemical equation.

Example. The combustion of ethane, C_2H_6,:

$$2C_2H_6 + 7O_2 \rightarrow 4CO_2 + 6H_2O$$

Suppose that we have on hand 352 g of ethane. How many moles of ethane do we have? How many moles of oxygen will be required to convert it to CO_2 and H_2O according to the recipe above? How many moles of CO_2 will enter the atmosphere?

Solution. The moles of ethane can be calculated from the given mass and the molar mass, which is 2(12.011) + 6(1.008) = 30.070 g/mole.

$$\text{moles ethane} = 352 \text{ g}/30.070 \text{ g/mole} = 11.71 \text{ moles}$$

According to the recipe, 7 moles of O_2 are required for each 2 moles of ethane consumed. Thus

$$\text{moles } O_2 \text{ required} = 7/2 \times \text{moles ethane}$$
$$= 40.97 \text{ moles oxygen.}$$

Again according to the equation, 4 moles of CO_2 are produced for each 2 moles of ethane burned:

$$\text{moles } CO_2 \text{ produced} = 4/2 \times \text{moles ethane}$$
$$= 23.42 \text{ moles carbon dioxide}$$

The type of calculation, based on the numerical relationships in the chemical equation, is extremely important in chemistry. We will practice these continually as we progress.

SUPPLEMENT: AN ALTERNATIVE APPROACH TO REACTION STOICHIOMETRY

Calculations important in chemistry, yet they seem difficult for many students. Here we present an alternative approach to doing this type of calculation that explicitly shows the amounts of substances present before reaction takes place; the amounts of substances that react and form; and the amounts present after reaction is complete. You may find this system easier to use than that outlined above. In addition, we will find this approach useful in our discussions of chemical equilibrium.

To illustrate the method, let us do a problem similar to that in Example:

The combustion of ethane, C_2H_6, is represented in 1-8-16:

$$2C_2H_6 + 7O_2 \rightarrow 4CO_2 + 6H_2O$$

Suppose that we have on hand 352 g of ethane and 450 g of oxygen.

How many moles of ethane do we have? How many moles of oxygen do we have? How many moles of CO_2 and H_2O can be produced according to the recipe above?

We begin as before with a balanced equation for the process:

$$2\ C_2H_6 + 7\ O_2(g) \rightarrow 4\ CO_2(g) + 6\ H_2O(l)$$

Directly under the equation, we write the *initial amount* (the amount we start with) of each substance:

	$2\ C_2H_6$ +	$7\ O_2(g) \rightarrow$	$4\ CO_2(g)$ +	$6\ H_2O(l)$
initial,g	352	450	0	0

Note that we explicitly acknowledge that before reaction takes place, we have no CO_2 or H_2O.

Knowing that the coefficients in the chemical equation represent the numbers of *molecules* or *moles* of substances involved, we first convert masses to moles, and portray the information in a second line labelled *initial, moles*:

	$2\ C_2H_6$ +	$7\ O_2(g) \rightarrow$	$4\ CO_2(g)$ +	$6\ H_2O(l)$
initial,g	352	450	0	0
initial, moles	11.71	14.06	0	0

Now we must decide which of the two reactants, ethane or oxygen, will be used up first. The equation helps us with this, because it indicates that for every 2 moles of ethane used, 7 moles of O_2 are needed. Thus to completely use up 11.71 moles of ethane would require $7/2 \times 11.71 = 40.99$ moles of O_2. Clearly there is not this much available initially, so we conclude oxygen is used up first; i.e., it is the *limiting reagent*. We acknowledge this by writing *lim* under the initial moles for oxygen:

	$2\ C_2H_6$ +	$7\ O_2(g) \rightarrow$	$4\ CO_2(g)$ +	$6\ H_2O(l)$
initial,g	352	450	0	0
initial, moles	11.71	14.06 lim	0	0

Next we write the amounts of reactants that react and the amounts of products that form.

These amounts are determined from the amount of the limiting reagent and the stoichiometric coefficients in the chemical equation:

	$2\ C_2H_6$ +	$7\ O_2(g) \rightarrow$	$4\ CO_2(g)$ +	$6\ H_2O(l)$
initial,g	352	450	0	0
initial, moles	11.71	14.06 lim	0	0
reacts/forms, moles	– (2/7)*14.06 = – 4.02	-14.06 lim	(4/7)*14.06 = 8.03	(6/7)*14.06 = 12.05

Note a couple of important things about the reacts/forms line. First, because reactants are used up as reaction takes place, the "reacts" amounts are entered as negative.

Second, the limiting reagent is assumed to completely react; i.e., to be used up. Third, the amount of ethane used up and the amounts of CO_2 and H_2O formed are calculated by applying the stoichiometric coefficients to the number of moles of limiting reagent.

Finally, the amounts of all substances present when reaction is over are obtained by adding the initial amounts to the amounts used up or formed:

	$2\ C_2H_6$ +	$7\ O_2(g) \rightarrow$	$4\ CO_2(g)$ +	$6\ H_2O(l)$
initial,g	352	450	0	0
initial, moles	11.71	14.06 lim	0	0
reacts/forms, moles	– (2/7)*14.06 = – 4.02	–14.06 lim	(4/7)*14.06 = 8.03	(6/7)*14.06 = 12.05
final, moles	7.69	0	8.03	12.05

If the final amounts are desired in grams, they may be easily calculated using molar masses:

	$2\ C_2H_6$ +	$7\ O_2(g) \rightarrow$	$4\ CO_2(g)$ +	$6\ H_2O(l)$
initial,g	352	450	0	0
initial, moles	11.71	14.06 lim	0	0
reacts/forms, moles	– (2/7)*14.06 = –4.02	-14.06 lim	(4/7)*14.06 = 8.03	(6/7)*14.06 = 12.05
final, moles	7.69	0	8.03	12.05
final, grams	30.1 g	0	353 g	217 g

Chapter 2

The Quantized Atom

THE DAWN OF QUANTUM THEORY

At the end of the 19th century, physics (hence the theory of chemistry) was considered nearly complete. The wave behaviour of electromagnetic radiation (light) was thought to be well understood in terms of the theory of James Clerk Maxwell, and a picture of the particulate atom was slowly emerging.

Only a few persistent but small problems remained to be cleared up. These have rather esoteric names: the problem of black body radiation; the photoelectric phenomenon; and the discrete emission and absorption spectra of atoms. Physicists thought these problems would ultimately be solved within the so-called classical (i.e., pre-Quantum) framework. Ultimately, however, this proved impossible.

Only by introducing the very strange and completely unfamiliar idea of quantized energy were they able to provide explanations for these rogue experiments.

In doing so, they changed forever not only the way that we think about the behaviour of matter at the atomic scale, but the way we think about measurement. In this text, it will not be appropriate for us to discuss the details of the explanations of the black body problem and the photoelectric effect. Suffice it to say that the idea of quantized energy was first introduced by Max Planck in 1900 in his explanation of the black body problem; and the idea of light bundles that we call photons was proposed by Einstein in 1905 to explain the photoelectric effect.

The Nature of Light

Before exploring in detail the so-called line spectra of atoms and explaining more fully what we mean by phrases like "quantized energy," we will briefly discuss the modern view of electromagnetic radiation—light—that has grown from the work of Maxwell, Planck and Einstein. Light has no mass, and is therefore not a form of matter. It is pure energy.

It travels very rapidly through space at a speed of 3.0×10^8 m/sec. This speed is so important that it is given a special symbol, c. According to the Theory of Relativity, the speed of light is the fastest possible speed, attainable only by massless things. Nothing can travel faster than this. We speak of light as if it consists of waves, somewhat like water waves. We assign to it a wavelength, l (lambda), and a frequency, n (nu), such that the product of the wavelength (a distance) and the frequency (a number of wavelengths per second that pass a given point) is the speed of light, c.

$$c = \lambda \nu$$

In fact, some behaviors of light are consistent with this picture. For example, light diffracts (passes through two neighboring holes in a screen and interferes with itself). This is a property usually associated with waves. However, other behaviors of light suggest that it is in some sense a particle. For example, shining light on the surface of a metal causes electrons of the metal to be ejected (this is the photoelectric effect).

The way in which electrons are ejected as a function of the frequency of the light suggests that each electron is ejected by being struck by a light packet. These light packets are called photons. The energy of a photon is directly proportional to its frequency, n:

$$E = h\nu$$

The proportionality constant, h, called Planck's constant, has the value 6.6×10^{-34} J-sec, where J stands for the energy unit, Joule (a Joule is the same as 1 kg-m^2/s^2). It carries Planck's name because he first introduced it in his explanation of black body radiation. This is another extremely important number in science, which we will encounter repeatedly. Apparently, then, light has both wave and particle characteristics, showing one face in some circumstances, the other face in other circumstances.

Equation explicitly recognizes the wave/particle duality of light by connecting energy, a particle-type quantity, with frequency, a wave quantity. The fact that light (and, as we shall see, electrons) shows both particle and wave behaviour is not a problem.

It simply means that our simplistic particle and wave ideas, which result from observing macroscopic (large scale) phenomena in the world around us, are inadequate for describing the behaviour of very small things.

Our modern view of light is as an electromagnetic oscillation in space. This view was developed by Maxwell in the 1860s and we still hold it today. Light waves are considered to be composed of sinusoidally oscillating, mutually perpendicular electric and magnetic fields. The frequency of the oscillation of both fields is the frequency of the light. An electromagnetic oscillation is shown in Figure.

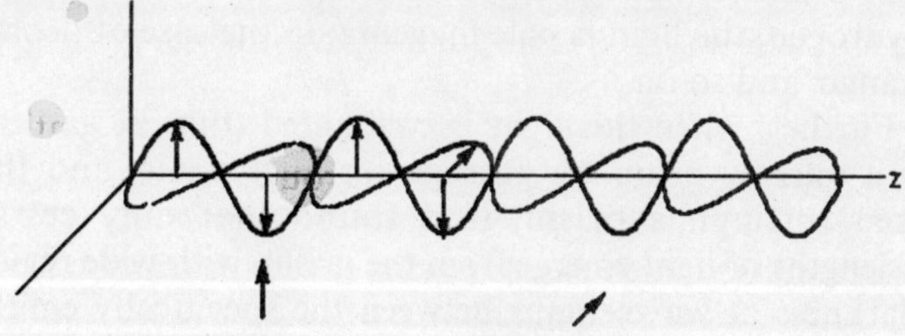

Fig. Electromagnetic Oscillation

Light, or electromagnetic radiation, exhibits a huge range of frequency (hence wavelength), from high energy (10^{20} s^{-1}, called *gamma* rays) to low energy (10^8 s^{-1}, radio waves). Light that is visible to the eye occupies a very narrow slice of this range. It is impossible for us to see most electromagnetic radiation! Light having a frequency somewhat less than we can see is called infrared (IR) radiation. That having frequency somewhat higher is ultraviolet (UV).

Example. The distance between the K^+ and Cl- ions in a crystal of KCl is 3.14×10^{-8} cm. What is the frequency of light whose wavelength will fit just once between the centers of the two ions?

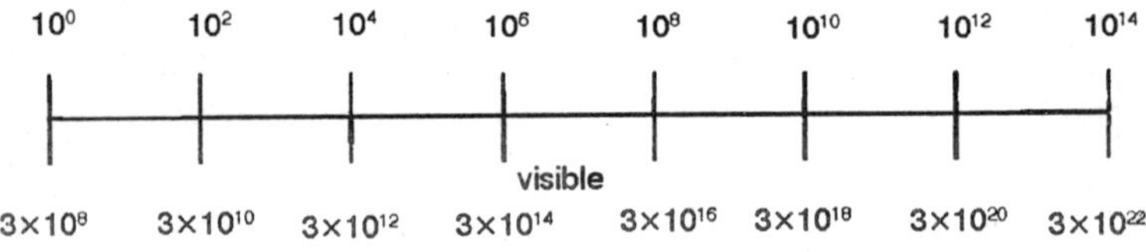

Fig. The Electromagnetic Spectrum

Solution. Use $c = \lambda\nu$

$$n = c/\lambda = (3.00\times10^{10}\ \text{cm/s})/(3.14\times10^{-8}\ \text{cm})$$
$$= 9.55\times10^{17}\ \text{s}^{-1}$$

This is in the X-ray region of the electromagnetic spectrum.

ATOMIC SPECTRA.

It was noticed as early as 1855 by Bunsen and Kirchoff that when atoms of a particular element, say hydrogen, are energized by heating or by electrical discharge, they emit (give off, or produce) light of a characteristic colour. In the case of hydrogen, the light is pale magenta; in the case of neon, it is orange; and so on.

Further, when this light is collimated (that is, gathered into a narrow beam by passage through a slit) and then passed through a prism, it is found that only certain wavelengths of light emerge from the prism, with wide regions of darkness at wavelengths between the specifically emitted values.

The collection of emitted lines is called the emission spectrum of the element. (It is also called a line spectrum, because the emitted beams take on the shape of the slit through which they are passed.) Figure shows the line spectrum of helium.

Most light sources (for example, the sun and stars, flames, and so on) produce a continuous spectrum of wavelengths when passed through a prism, just like the rainbow produced by passage of sunlight through mist or rain. That is, all wavelengths in the visible region of the spectrum are produced, with none missing. The discrete line spectra of atoms were therefore considered unusual, and could not be readily explained.

Fig. Lline Spectrum of Helium

The discrete emission spectra of certain atoms meshed nicely with an observation that had been made in 1814 by Fraunhofer. Using a high quality prism, he discovered that the spectrum of light from the sun, although essentially continuous, contains a number of dark (black) lines representing wavelengths where no light is seen. The positions of these dark lines matched exactly the positions of lines in the emission spectra of a number of elements, notably hydrogen, helium, and sodium.

The eventual interpretation of the Fraunhofer lines is that they result from absorption of light at the dark wavelengths by elements present in the outer portion of the gaseous cloud of the sun. The Fraunhofer lines constitute the absorption spectrum of these elements. The absorption and emission spectra of an element are complementary. The absorption spectrum shows dark lines against a bright background. The dark lines are at wavelengths where light is absorbed, and thus not seen. The emission spectrum shows bright lines against a dark background. The positions of the lines in the two types of spectra coincide exactly. Line spectra were studied for many years without adequate explanation. A number of scientists focussed on the spectrum of hydrogen, which appeared the simplest. The series of lines falling in the visible region, called the Balmer series.

The series is so named for Jacob Balmer, who found that the reciprocal wavelengths of the lines obey equation.

$$1/\lambda = R_H(1/2^2 - 1/n^2)$$

R_H is a constant with value 1.09678×10^{-2} nm^{-1}. Note that reciprocal wavelength is closely related to frequency, hence energy, via equations. Balmer was unable to explain the meaning of the integers occuring in the equation.

Nonetheless, their occurence, and the very simple form of equation, are remarkable.

By recording the emission spectrum of hydrogen on light sensitive film, scientists were able to "see" not only the visible emission lines, but also emission lines in the ultraviolet and infrared regions of the electromagnetic spectrum. Thus between 1906 and 1924, four additional series of lines were discovered. All of the observed spectral series of the hydrogen atom were found to obey a generalized version called the Rydberg equation below:

$$1/\lambda = R_H(1/n_1^2 - 1/n_2^2)$$

The unified fit of the hydrogen spectrum provided by equation suggested a regular internal structure to the atom that is changed in some way by interaction with light. Nothing was known about this internal structure in 1885, when Balmer proposed his equation. In fact, the knowledge that atoms are composed of still smaller and more fundamental units of matter was not known. The origin of the emitted wavelengths of light within the atom was therefore not understood.

The first definite knowledge of the substructure of the atom emerged from the experiments of JJ Thomson in 1897. Using a device known as a cathode ray tube, Thomson demonstrated that the "cathode ray" consists of a beam of negatively charged particles that emerge from the metal cathode of the tube under an applied voltage. He called these particles electrons. By observing the manner in which the beam deflected when subjected to magnetic and electric fields of known strengths, Thomson was able to calculate a value of about 10^8 for the charge-to-mass ratio of the electron. The modern value is given in equation:

$$\text{charge/mass} = e/m = 1.76\times10^8 \text{ C/g}$$

Subsequent experiments by Robert Millikan between 1908 and 1917 provided the value 1.6×10^{-19} coulombs for the charge, leading to a mass of 9.11×10^{-28} g.

Subsequent to Thomson's work, the atom was viewed as a uniform distribution of electrons embedded in a uniform distribution of positive charge, required to be present to ensure the electrical neutrality of the atom. However, in 1911 Rutherford published the so-called nuclear model of the atom, based on the results of experiments in which he and his students bombarded thin gold foil with alpha particles, which are the nuclei of helium atoms. The results of these experiments were interpreted by Rutherford as indicating that the positive charge of the atom was concentrated in a tiny and extremely dense region at the centre of the atom, which he called the nucleus. He proposed that virtually all of the mass of the atom resides in the nucleus. The electrons were then proposed to orbit the nucleus, much as the planets orbit the sun.

BOHR'S THEORY OF THE HYDROGEN ATOM

In 1913, Niels Bohr, a young Danish physicist, provided the first successful theory of the internal structure of the hydrogen atom, based on the quantum ideas of Planck and Einstein. He conceived his theory by speculating about the origin and significance of atomic line spectra.

According to Bohr's interpretation of atomic spectra, the absorbed and emitted photons provide a direct measure of the gaps between discrete (quantized) energy levels in the atom.

To arrive at a quantitative expression for the energies of these so-called energy levels in the hydrogen atom, Bohr made several assumptions, some quite arbitrary (i.e., the assumptions were justified only because they ultimately allowed an explanation of line spectra).

- The electron moves about the nucleus in well-defined circular orbits. Only certain orbits, each with a definite angular momentum, are available to the electron. The angular momenta of the electron in these orbits are integer multiples of Planck's constant, h, divided by 2p:

  $$nh/2p = mvr$$

  Here m and v are the mass of the electron and its velocity in the orbit, and r is the radius of the orbit. n is an integer, which Bohr called a quantum number, with value 1 in the smallest orbit, 2 in the next larger orbit, and so on.

- Each orbit is associated with a particular energy for the electron, with energy increasing with the radius of the orbit. The energy of the electron in the hydrogen atom is therefore quantized; that is, it is allowed to have only certain values. Each allowed value characterizes an energy level of the atom.
- The atom absorbs or emits a light photon when the electron changes from one energy level to another within the atom. The electron can be made to move to a higher energy level by absorption of a photon; when the electron moves to a lower energy level, a photon is emitted by the atom.
- The energy of the absorbed or emitted photon exactly matches the difference in energy between the final and initial energy levels of the electron:

$$:hn = E_f - E_i$$

The key assumption made by Bohr is the first one, which restricts the electron to certain discrete locations within the atom, while forbidding all others. In proposing this, Bohr broke with classical electromagnetic theory, according to which an electron orbiting a positive nucleus would continuously radiate its energy away, eventually spiralling in to the nucleus.

With the first assumption, Bohr superimposed Planck's quantum ideas on the electron in the atom. In restricting the momentum of the electron, he restricted in turn not only the radii of the orbits, but also their energies. Assuming that the attractive force exerted on the electron by the nucleus, given by Coulomb's Law, was balanced by the centrifugal force due to the electron's orbital motion, Bohr was able to show that the orbit energies are given by equation:

$$E_n = -me^4/8h^2 e^2n^2 = -2.176\times10^{-18}/n^2 \text{ Joules}$$

In equation above, e is the electron charge, with value 1.602×10^{-19} coulombs; and e is a constant of nature called the vacuum permittivity, with value 8.85×10^{-12} $J^{-1}C^2M^{-1}$. We will not be concerned with the significance of the vacuum permittivity, other than to recognize that it causes the units to work out properly.

The important feature of equation is that it shows a direct dependence of the electron energy on the quantum number n, which may have only integer values. Thus only certain energies can occur! The atom is quantized.

Example. The idea of quantization is not entirely unfamiliar at the macroscopic level. What are some examples of things that are quantized?

Solution. Eggs, apples, oranges

Stair steps

Floors in housesCurrency

Equation is seemingly strange in one respect—it states that the energy of the electron in the hydrogen atom is negative! What do we take this to mean? It seems reasonable to think of the energy of the electron as associated somehow with its motion. There are two types of energy. The first is kinetic energy, energy of motion, given by the expression below.

$$KE = 1/2\ mv^2$$

Here m is the mass and v the velocity of a particle or object (this equation shows the relationship given earlier between the energy unit, Joule, and units of mass and velocity). The second is potential energy, energy of position. The electron in the hydrogen atom has energy of both types. It is of course not possible for kinetic energy to be negative, because the concept of negative motion makes no sense.

The minimum possible value for kinetic energy is zero, corresponding to no motion. However, potential energy is often negative. The familiar example of potential energy is gravitational potential energy, associated with the position of some object in the earth's gravitational field. We normally take ground level to be the zero point of potential energy.

That is, when an object, say a basketball, is sitting on the surface of the earth, its potential energy is taken to be zero. This is done for our convenience. When we raise the basketball to a higher position, we do work on it, and its potential energy increases (becomes positive). However, if the basketball were to fall into a hole in the ground, its potential energy would decrease—it would become negative—because we would have to do work on the basketball to raise it out of the hole and bring its potential energy back to zero.

Thus potential energy can be either positive or negative, depending on the position of the system with respect to the agreed upon zero-point position. The electron has potential energy by virtue of the force of attraction to the positive nucleus according to Coulomb's Law:

$$F = ke^2/r^2$$

The potential energy is electrical, rather than gravitational, but it is mathematically similar. The closer the electron gets to the nucleus, the lower its potential energy becomes, just as an object in the earth's gravitational field decreases in potential energy the closer it gets to the centre of the earth. By agreement among scientists, the potential energy of the electron-proton system is taken to be zero when the force of attraction is zero. According to equation, this will happen when r = infinity; that is, when the electron and proton are infinitely separated.

When the electron is pulled away from the nucleus to an infinite distance, the potential energy of the electron is zero. At any closer distance than this, the potential energy is negative. As it turns out, the potential energy of the electron in the hydrogen atom has larger magnitude than the kinetic energy. Since the potential energy is negative, the total energy, given by equation, is negative.

(To give a simple example, if the potential energy were -5 kJ and the kinetic energy were 2.5 kJ, the total energy would be -2.5 kJ.). The fact that the energy of the electron in the H atom is negative is thus not problematic. It is a simple consequence of our definition of potential energy for the electron-nucleus system. We will find that electron energies in atoms are always negative, due to the potential energy contribution to the total energy.

The triumph of the Bohr theory results from applying to electron transitions—processes in which the electron moves from one allowed energy level to another, with the accompanying absorption or emission of a photon. Suppose that the electron is initially in the orbit with n = n_i. It absorbs a photon of energy hn and moves out to the orbit with n = n_f. According to Bohr, there must be an exact match of the photon energy and the difference in energy between orbits n_i and n_f, according to equation. Thus $hn = -(me^4/8h^2e^2)(1/n_f^2 - 1/n_i^2)$

Substituting c/l for n and dividing both sides by hc gives the expression for $1/\lambda$:

$$1/\lambda = -(me^4/8h^3c\ e^2)(1/n_f^2 - 1/n_i^2) = -C((1/n_f^2 - 1/n_i^2)$$

This equation is identical in form to equation, which describes the dependence of the wavelengths of the emission lines of hydrogen on integers of unknown significance. When Bohr calculated the value of his constant, C, from known values of the mass and charge of the electron, the vacuum permittivity,

Planck's constant, and the speed of light, he obtained the value 1.09678×10^{-2} nm^{-1}. This agrees so closely with the experimentally determined value of the Rydberg constant, R_H, that Bohr's theory of hydrogen was immediately and enthusiastically accepted, despite its flagrant break with the classical view. The allowed energy levels of the hydrogen atom according to equation. Such a plot is usually called an energy level diagram. Transitions between levels are shown as arrows that point from the initial to the final level.

The lines in the emission spectrum of hydrogen are due to photons that are emitted by the atom when the electron moves from a higher to a lower level. Arrows corresponding to several lines in the Balmer emission series, in which the electron moves from levels with n > 3 to the n = 2 level, are shown.

The dark lines in the absorption spectrum of hydrogen are due to the absence of photons that are absorbed by the atom, causing the electron to move from a lower to a higher level. Arrows corresponding to the Balmer absorption series are also shown. Note that the difference in energy between two levels is fixed by equation. There is thus an exact correspondence in energy of the emission and absorption lines.

Example. What is the frequency of the photon required to promote the electron from the n = 1 level to the n = 4 level in the hydrogen atom?

Solution. According to equation,

$$E = 2.176\times10^{-18}(1/n_1^{\ 2}-1/n_2^{\ 2})$$

The energy of the required photon is $2.176\times10^{-18}(1-1/16) = 2.04\times10^{-18}$ J. Since E = hn, n = E/h = 2.04×10^{-18} J/ 6.626×10^{-34} J-s = 3.08×10^{15} s^{-1}

After his initial success with the hydrogen atom, Bohr attempted to apply his theory to the helium atom, which has two electrons orbiting a nucleus consisting of 2 protons and 2 neutrons. Despite years of effort and numerous modifications to his theory, Bohr was unable to extend his success beyond the hydrogen atom. Nonetheless, Bohr had taken the first step in the development of the modern theory of matter and energy on the microscopic scale—Quantum Mechanics. The second breakthrough step was made by Louis de Broglie in 1924.

WAVE PARTICLE DUALITY—MODERN QUANTUM THEORY

In perhaps one of the most famous and significant Ph.D theses in all of science, Louis de Broglie laid the foundation for further advances in the quantum ideas. He suggested that matter, like light, might exhibit a duality of wave and particle characteristics. In other words, he suggested that the electron, normally considered a particle, might behave sometimes as a wave! His suggestion was purely intuitive.

There was absolutely no experimental evidence at the time to warrant it. Nonetheless, he was able to put it on a quantitative basis by making an analogy to a photon, for which the relationship between wavelength and momentum, p, is given in equation: l = h/p de Broglie suggested that the electron is governed by a similar equation. Substituting the product of mass and velocity for momentum, he obtained equation:

$$l = h/mv$$

He then proceeded to show that Bohr's theory of the hydrogen atom was a consequence of this equation! His reasoning was (probably) something as follows. The allowed orbits of the hydrogen atom are those in which an integral number of electron wavelengths exactly fits the circumference of the orbit. In other words, de Broglie viewed the electron in the hydrogen atom as a standing wave, analogous to the standing waves set up in a guitar string when it is plucked. Just as the number of wavelengths must exactly fit the length of the guitar string, which is tied down at both ends, so must the electron wave fit the orbit. If this were not true, the electron wave would interfere with itself, and the electron would disappear. In quantitative terms, de Broglie's assumption is stated as in equation:

$$nl = 2pr$$

Here n is an integer and r is the orbit radius. Substituting for the electron wavelength from equation:

$$nh/2p = mvr$$

This is exactly Bohr's quantized momentum assumption, from which all the results of Bohr's theory follow. It is important to realise that although de Broglie's result for the hydrogen atom is the same as Bohr's, de Broglie's approach is more satisfying, because the quantum number n arises quite naturally from the requirement that the electron be a standing wave that just fits the circumference of the orbit.

This contrasts markedly with the necessity for Bohr to superimpose the quantum condition in ad hoc fashion on an otherwise quite classical picture of the hydrogen atom. The ideas of de Broglie were not taken completely seriously at first, because the idea of the electron as a wave was difficult to accept for many scientists. Further, there was no experimental support for his ideas. However, in 1927, Davisson and Germer showed that when beams of electrons are made to impinge on a crystalline material, a diffraction pattern results.

Diffraction is a property of waves, not of particles. Further, the electron wavelength extracted by Davisson and Germer from their results was just the value expected from de Broglie's relationship, equation. The wave nature of the electron was thus firmly established in experiment.

Example. The electron in the n = 1 level of the hydrogen atom travels at 2.19×10^6 m/s. What wavelength is associated with the electron? How many such wavelengths fit in the first Bohr orbit of hydrogen (radius of orbit = 0.529×10^{-8} cm).

Solution. We use the de Broglie relation to calculate the wavelength, then divide this by the circumference of the orbit:

l = h/mv = 6.626×10^{-34} J-sec/(9.11×10^{-31} kg)(2.19×10^6 m/s) = 3.32×10^{-10}m = 3.32×10^{-8} cm.Circumference of orbit = 2pr = 2(3.1416)(0.529×10^{-8}) cm = 3.32×10^{-8} cm

Thus l/2pr = 1 ! Exactly one electron wavelength fits around the first Bohr orbit.

It is interesting and revealing to cast equation in somewhat different form. Reciprocating both sides of gives:

$$1/\lambda = mv/h = p/h$$

We recognize that $1/\lambda$ is the number of wave cycles that occur per unit length (m or cm). For this reason, it is usually called the wavenumber, and is given the symbol k. Wavenumber is similar in nature to frequency, which is the number of cycles that occur per unit time. Replacing 1/l with k and multiplying both sides by h gives: p = hk

Clearly this equation is very similar in form to equation. Like that equation, it relates a particle-type quantity (momentum) to a wave-type quantity (wavenumber) through the quantum constant h. The de Broglie relation expressed as equation thus also explicitly recognizes the wave-particle duality of matter, just as equation recognizes that of light.

The Final Step—the Schrodinger Wave Equation.

The final stone in the foundation of modern quantum mechanics was laid by Erwin Schrodinger in 1926. He realized that if the electron were indeed a wave, as de Broglie claimed, it should be governed by a mathematical relationship analogous to the equations that describe ordinary waves at the macroscopic scale.

He set out to find such a relationship. The result of his effort, the Schrodinger equation, is the basis for our understanding of the quantum characteristics not only of the hydrogen atom, but of all atoms and molecules:

$$(2-4-7): -(h^2/8\pi^2 m)(\partial^2\psi/\partial x^2 + \partial^2\psi/\partial y^2 + \partial^2\psi/\partial z^2) + V_\psi = E\psi$$

This intimidating-looking equation is called a second-order differential equation.

The same type of equation governs three-dimensional wave motions of all types. We will not be at all concerned with the manner in which this equation was obtained, nor with how it is used or solved.

That will be left for higher level texts that are concerned with specialized areas of chemistry. However, we will be interested, at least qualitatively, in the solutions to this equation, particularly those for the hydrogen atom. It is to the hydrogen atom that Schrodinger first applied his new equation, with resounding success.

We now look at his results, in a simplified way. We must first discuss the various terms and symbols in equation. The symbols h and m have the same meaning as in Bohr's equation: they are Planck's quantum constant and the mass of the electron.

The symbol, ∂, is a partial derivative symbol, used in calculus, and will not concern us now. V is the potential energy of the electron in the electric field of the nucleus. We discussed this in talking about the Bohr Theory. E is the total energy of the electron (potential plus kinetic); and y is the wave function of the electron. We are now in a position to summarize the major results of Schrodinger's solution for the hydrogen atom.

- Solution of the equation gives rise to 3 quantum numbers. The first is n, which is equivalent to the quantum number n in Bohr's Theory. The remaining two are given the symbols *l* and m_l. Three quantum numbers rather than one arise because the electron in the H atom is capable of motion in 3 dimensions (x, y, and z in the Cartesian coordinate system). The standing wave pattern of the electron must therefore be restricted by a quantum number in each of these 3 dimensions. In general in quantum mechanics, there is a quantum number required for each direction of motion available to the system. As in the Bohr theory, quantum numbers are restricted to integer values. The values allowed for n, l, and m_l are described below:

– n, the principal quantum number, has integer values ranging from 1 to infinity. Each value of n is said to characterize a quantum shell, or main shell, in which the electron could possibly reside.
– l has integer values ranging from 0 to (n-1). Thus when n = 1, only $l = 0$ is possible. When n = 2, l can be 0 or 1, and so on. Each value of l for a particular value of n is said to characterize a subshell of the main shell corresponding to n. Thus for the n = 1 shell, there is a single subshell (l = 0); for the n= 2 main shell, there are two subshells (l = 0 and 1), and so on. For historical reasons, the various values of l are assigned letter designations, as follows:

  l = 0 1 2 3l is s p d f

  The 2s subshell is therefore the subshell with l = 0 in the n = 2 main shell. The 3d subshell has n = 3 and l = 2; and so on.
– m_l has integer values ranging from $-l$ to $+l$. Thus for $l = 0$, only $m_l = 0$ is possible; for $l = 1$, m_l can be -1, 0, or +1; and so on. Each value of m_l for a particular value of l is said to characterize an orbital of the subshell. The orbitals belonging to, say, the 2p subshell are collectively called the 2p orbitals.

  Normally the electron in the hydrogen atom is found in the orbital with m_l = 0 in the 1s subshell. This is the orbital of lowest (most negative) energy. The electron can be promoted to orbitals in higher energy shells by absorption of light.

  Table. Summarizes the Main shell/Subshell/ Orbital Hierarchy of the Hydrogen Atom

Table. Shell Structure of the Hydrogen Atom

n (main shell)	*l (subshell)*	m_l *(orbital)*
1	0	0
2	0	0
2	1	–, 0, 1
3	0	0
3	1	–1, 0, 1
3	2	–2, –1, 0 1, 2
4	0	0
4	1	–1, 0, 1
4	2	–2, –1, 0, 1, 2
4	3	–3, –2, –1, 0, 1, 2, 3

- The Schrodinger equation gives the allowed values of total energy for the electron in the H atom. The energy depends on the quantum number, n, according to the same equation obtained by Bohr:

$$E_n = -me^4/8h^2e^2n^2$$

As seen above, the quantum number, n, has the same values and much the same meaning as in Bohr's Theory. However, in Schrodinger's theory, the quantum number arises naturally, as in de Broglie's theory, from the standing wave assumption.

- Solution of the Schrodinger equation gives mathematical expressions for the wave functions, y, associated with each of the allowed orbitals of the hydrogen atom. Although Schrodinger was able to obtain these wave functions readily, he was uncertain as to what they meant.

His colleague, Max Born, suggested that the square of the wave function, y^2,in a particular region of space should be interpreted as the probablility of finding the electron in that region of space. Although Schrodinger was not satisfied with this interpretation, many other scientists were. This interpretation has become the so-called orthodox interpretation of the wave function. Although there are alternative interpretations, the orthodox interpretation is accepted and taught by most scientists today, and we will adopt this view of the wave function in this text.

A visual presentation of the location-probability of the electron in a particular orbital of the atom can be obtained by constructing plots of the mathematical expression for the orbital wave function. Two types of plots are typically made. The first shows the manner in which the location probability varies with distance of the electron from the nucleus of the hydrogen atom. These are called radial plots. Radial plots for the 1s, 2s, 2p, 3s, and 3d orbitals of the hydrogen atom are shown in Figure. Several features of these plots should be noted.

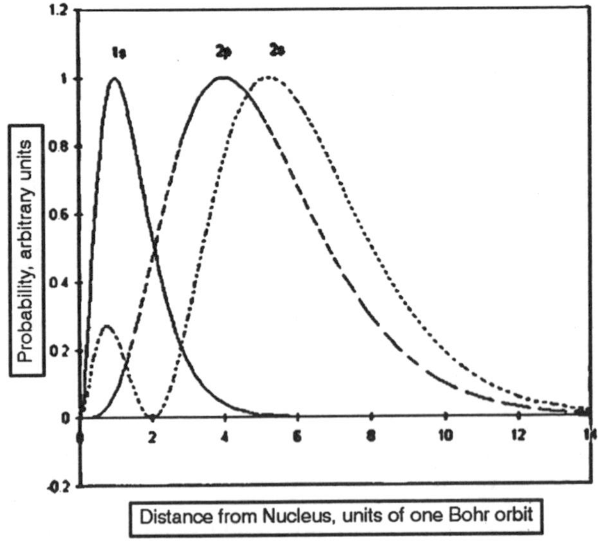

Fig. Radial Plots for Hydrogen Atom Orbitals

First, there is no single particular distance at which the electron is located with respect to the nucleus in any orbital. This is in marked contrast to the Bohr theory, in which electrons are constrained to fixed circular orbits of definite radius. According to the modern view, the electron in, say, the 2s orbital can be found at any distance from the nucleus, although the probability is highest of finding it at a distance of 2.116×10^{-8} m. This corresponds to the maximum in the curve. Second, as the value of n increases, the maximum in the plot moves to larger distance, meaning that the electron tends to be further from the nucleus in shells with larger n (higher energy). This is qualitatively similar to the results of Bohr's Theory, in which the radii of allowed orbits became larger with increasing n.

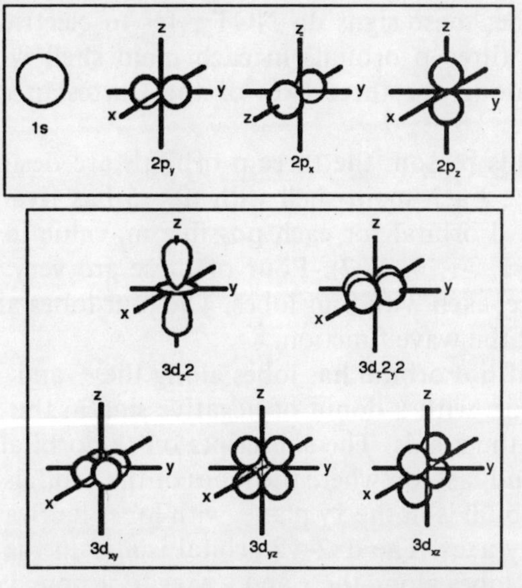

Fig. Angular Plots for Hydrogen Orbitals

Third, the radial plot for the 2s orbital shows that there is zero probability of finding the electron at 1.06×10^{-8} m from the nucleus, because the radial plot goes to zero here. This region of zero probability is called a node. It is analogous to the nodes in a standing wave in a guitar string, which are locations at which the string does not move. Nodes are characteristic of wave patterns; their occurrence in the wave functions for the electron in the H atom reinforces the de Broglie picture of the electron as a wave.

The second type of plot shows the 3-dimensional shape of the orbital. These are usually referred to as angular plots. Angular plots for the 1s, 2s, 2p, and 3d orbitals are shown. Again, there are a number of important features of these plots. First, the angular plots for s orbitals are spherical in shape. This implies that an electron in such an orbital can be found with equal probability in any direction from the nucleus. In contrast, p orbitals are dumbbell-shaped. Each of the sphere-like portions of a p orbital is called a lobe. p orbitals are therefore said to have 2 lobes. The + and–signs in the orbital lobes refer to the sign of the wavefunction, y, in that lobe; these signs do NOT refer to electrical charge. There are three p orbitals in each main shell with n > 2, oriented along the three axes of the Cartesian coordinate system.

For this reason, the three p orbitals are designated px, py, and pz. Each main shell with n > 3 has five d orbitals (there is a d orbital for each possible m_l value in the $l = 2$ subshell: –2, –1, 0, 1, 2). Four of these are very similar in appearance, each with four lobes. The four lobes alternate in the sign of the wave function.

The fifth d orbital has lobes along the + and–directions of the z axis, with a donut of negative sign in the x-y plane, encircling the z axis. The subscripts on the orbitals indicate in shorthand fashion where the lobes of the orbitals are. Thus the dxy orbital is in the xy plane, with lobes halfway between the x and y axes. The dx2–y2 orbital is also in the xy plane, but has its lobes along the x and y axes. For now, you should familiarize yourself with the shapes of s and p-type orbitals.

THE HEISENBERG UNCERTAINTY PRINCIPLE

It may puzzle you that we have changed rather abruptly from discussing exact electron orbits in the Bohr theory to rather fuzzy probabilities in the modern quantum theory. Indeed, one of the remarkable things that we have learned from quantum theory is that we can not know the exact location of the electron in the hydrogen atom (or of any electron in any atom). We can know at best only the probability that the electron will be found in a particular region.

This lack of exact knowledge is a consequence of one of the strangest and most fundamental principles of quantum mechanics, the Uncertainty Principle, proposed by Werner Heisenberg in 1927. There are numerous ways to state this principle, one of which follows: it is not possible to simultaneously know exactly both the position and the momentum of an electron (or of any other small "particle"). As we have seen, the momentum of a particle-wave is the product of its mass and velocity.

We may know the position exactly, but can then have no idea what the momentum is; similarly, if we know the momentum exactly, we can have no idea what the position is. This is difficult to accept because at the macroscopic scale, we of course are able to know both quantities at the same time. For example, when you drive your car to the grocery store, you know how fast you are going (momentum) and you know exactly where you are at each moment (position). Very small particles simply do not behave in the same way as large ones. This is not something that can be explained; it simply is.

Just as Heisenberg's principle links momentum and position in uncertainty, it also links energy and time. In more mathematical terms, Heisenberg's Uncertainty relations are expressed as in equations 2-4-8.

$$(Dp) \times (Dx) > h/2p$$
$$(DE) \times (Dt) > h/2p$$

In words, these twin relationships say that the product of the uncertainty (indicated by D) in momentum (or energy) and the uncertainty in position (time) must exceed a fundamental minimum value given by h/2p. This minimum value is so small that we do not observe the Uncertainty Principle working at the macroscopic scale.

In the atomic/molecular realm, however, it is extremely important. (For those with an interest in the quantum view of nature, there is a third Heisenburg relationship involving the uncertainties in angular momentum, L, and angular position, w, written $(DL)\times(Dw) > h/2p$. This relation is less-used than the two above and will not concern us.)

Example. An electron is accelerated through a potential difference of 1 volt, which causes it to acquire kinetic energy. The maximum kinetic energy that the electron could acquire is 1 electron-volt, or 1.6×10^{-19} joules. Calculate the uncertainty in the position of the electron.

Solution. The kinetic energy of the electron is somewhere between 0 and 1.6×10^{-19} joules. Let's calculate the momentum of a 1.6×10^{-19} joule electron. This will be the maximum possible momentum. The minimum momentum possible is 0.

$$KE = 1/2\ mv^2 p = mv$$

Thus $p^2 = 2mKE$, and $p = (2mKE)^{1/2}$. We calculate that $p = 5.40\times10^{-25}$ kg-m/s. Because the momentum lies somewhere between 0 and this value, the uncertainty in p is roughly 10^{-25} kg-m/s. Applying the Heisenberg principle, the uncertainty in position is

$$Dx = h/2pDp = 6.6\times10^{-34}\ \text{joule-s}/(2p)$$
$$(10^{-25}\ \text{kg-m/s}) = 1.0\times10^{-9}\ \text{m}$$

As we have stated earlier, atoms are not possible in terms of classicial physics, according to which the electrons should be attracted into the nucleus, radiating their energy away as they fall.

There is no doubt whatsoever, though, that this does not happen. Electrons occupy regions of space called orbitals, with definite energies. The Uncertainty Principle allows us to understand the existence of atoms. If electrons were to spiral into the nucleus, losing all of their energy, then we would simultaneously know where they were (in the nucleus) and what their momentum is (zero).

This would violate the Uncertainty Principle. The electron is a standing wave in the atom, with a definite wavelength, and consequently a definite momentum. The Uncertainty Principle therefore requires that we have no knowledge about where it is! In fact, the radial plots in Figure above show that the electron can be found with some probability at any distance (even an infinite one) from the nucleus; at any given instant we do not know where it is. The Uncertainty Principle is extremely important—it allows the existence of atoms.

With the Uncertainty Principle in mind, it is instructive to return to the radial plots for the orbitals of the hydrogen atom. Figure shows an enlarged version of the radial plot for the 2s orbital. What does the plot tell us about the electron, and in particular about its distance from the nucleus? First, the radial plot has the visual appearance of a wave; it oscillates up and down, with clear peaks and troughs.

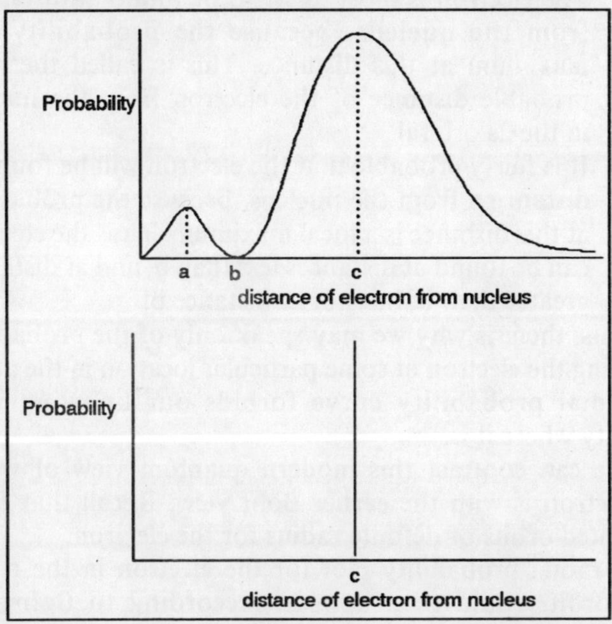

Fig. Radial Plot for the 2s Orbital According to (a Schrodinger b) Bohr

This is certainly consistent with the de Broglie/ Schrodinger view of the electron as a wave. In addition, it reveals the following specific information about where the electron is and is not:

- The electron is not at the nucleus; the probability is zero at a distance of zero.
- The electron is not at distance b from the nucleus; the probability is zero at this distance, too, meaning that the electron is never at that specific distance from the nucleus.
- The electron may be at any distance from the nucleus other than distance 0 and distance b. In fact, the curve approaches zero probability asymptotically as the distance becomes very large, so there is a small

but finite probability of finding the electron very far away from the nucleus.

- The electron is most likely to be found at distance c from the nucleus, because the probability is a maximum at this distance. This is called the most probable distance of the electron from the nucleus in the 2s orbital.
- It is fairly probable that the electron will be found at distance a from the nucleus, because the probability at this distance is a local maximum. Thus the electron can be found at distances less than b, and at distances greater than b, but not at distance b!

This, then, is why we may speak only of the probability of finding the electron at some particular location in the atom: the radial probability curve forbids our knowing with certainty where it is.

We can contrast this modern quantum view of where the electron is with the earlier Bohr view. Recall that Bohr postulated orbits of definite radius for the electron.

A radial probability plot for the electron in the n = 2 Bohr orbit would then appear. According to Bohr, the probability of finding the electron is zero everywhere except at distance c. Thus the plot rises to probability = 1 at distance = c, and has the value zero everywhere else. It is interesting that the radius, c, of the Bohr orbit for n = 2 is exactly the same as the most probable distance predicted by Quantum Theory for the 2s orbital: 2.116×10^{-10} m.

ATOMS WITH MORE THAN ONE ELECTRON ELECTRON SPIN AND THE PAULI EXCLUSION PRINCIPLE

The line spectra of many atoms other than hydrogen were well-known by the time Schrodinger developed his equation. In general, these spectra were much more complex than that of hydrogen. Even the relatively simple helium atom, with only 2 electrons, demonstrated a more complex spectrum than hydrogen.

It was understood that the complexity was due to the

presence of more than one electron. It was also realized that, contrary to what might be expected, these electrons did not all occupy the lowest energy Bohr-type shell.

If this were the case, then all atomic spectra would show the same simple pattern of line spacings (although not the same wavelengths) of the hydrogen spectrum. When Schrodinger presented his quantum mechanical view of the hydrogen atom, it was recognized that shells, subshells, and orbitals similar to those in hydrogen must exist in all atoms, and that electrons must occupy these according to certain rules. In 1925, Wolfgang Pauli proposed the most important of these rules, called the Exclusion Principle:

At most 2 electrons may simultaneously occupy an atomic orbital.

Prior to the discovery of the Schrodinger wave equation, improvement in the quality of instruments for measuring atomic spectra had shown that what had previously appeared to be single lines in the spectra of certain elements were in fact doublets—two closely spaced lines! In 1925 it was proposed that the doubling of lines was a consequence of a previously unsuspected type of electron motion, electron spin. The spin of the electron was proposed to be similar to the rotation of the earth on its north-south axis as it orbits the sun.

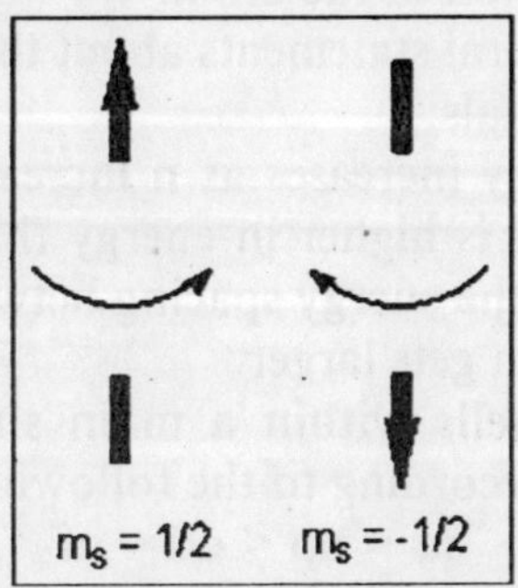

Fig. Electron Spin

(Note that this description of electron spin must be taken with a grain of salt; we have learned that we must think of the electron in the atom as a wave, not a particle. What does it mean for a wave to be "spinning on its axis"?) The quantum number associated with this motion was dubbed the electron spin quantum number, and was given the symbol m_s.

In order to explain the doubling of lines in atomic spectra, it was necessary to restrict this quantum number to only two values: +1/2 and -1/2. These two values are referred to as "up spin" and "down spin", respectively. This new quantum number is unusual in 2 respects. First, it can have only 2 values. Second, the allowed values are not integers. Instead, they are half integers.

The existence of the spin quantum number, with only 2 values, suggests an explanation for the Pauli Principle. Two electrons in the same orbital must have the same values of n, *l*, and m_l. However, If they have opposite spins (one "up", the other "down"), then they will differ in at least 1 of the 4 quantum numbers that characterize them. A restatement of the Pauli Principle in these terms is as follows:

Two electrons in an atom must differ in at least one of the four quantum numbers.

Electron Configurations. Armed with the Pauli Principle and the atomic orbitals from the Schrodinger treatment, we are in a position, for any atom, to specify the distribution of electrons among the orbitals. This distribution is called the electron configuration of the atom. We will take on faith (for the time being) several statements about the energies of shells, subshells, and orbitals:

- Shell energy increases as n increases. Thus the shell with n = 2 is higher in energy than the n = 1 shell. However, the energy spacing between shells becomes smaller as n gets larger.
- The subshells within a main shell have different energies, according to the following ordering.

  $$ns < np < nd < nf$$

  (Of course, for n = 1 there are no p, d, or f subshells; for n = 2 there are no d and f subshells; for n = 3 there is no f subshell.)
- A consequence of the first two statements is that for n^{3} 3, there is some overlap of the higher energy

subshells of a main shell with the lower energy subshells of the next higher main shell.

- The orbitals within a subshell all have exactly the same energy. Orbitals with the same energy are said to be degenerate.
- The maximum number of electrons that can enter a subshell is determined by the 2-per-orbital rule:

s subshell	2 Ectrons
p	6 Eectrons (3 orbitals, 2 per orbital)
d	10 Eectrons (5 orbitals, 2 per orbital)
f	14 Eectrons (7 orbitals, 2 per orbital)

The order of subshell filling consistent with these statements is the following:

1s 2s 2p 3s 3p 4s 3d 4p 5s 4d 5p 6s 4f 5d 6p 7s 5f (6d 7p) (2–5–1)

Note that there is some overlap of the high energy subshell of n = 3, 3d, with the low-energy subshell of n = 4, 4s. Similar overlaps occur more frequently as n increases.

Example. What is the electron configuration for each atom: B, Ca, Fe, Bi?

Solution. All we need for each atom is the number of electrons—the atomic number. We then distribute these electrons according to the order of filling and the allowed subshell populations. B 5Ca 20Fe 26Bi 83

The electron configurations are as follows. The number of electrons occupying a set of orbitals (a subshell) is indicated as a superscript:

B $1s^2 2s^2 2p^1$

Ca $1s^2 2s^2 2p^6 3s^2 3p^6 4s^2$

Fe $1s^2 2s^2 2p^6 3s^2 3p^6 4s^2 3d^6$

Bi $1s^2 2s^2 2p^6 3s^2 3p^6 4s^2 3d^{10} 4p^6 5s^2 4d^{10} 5p^6 6s^2 4f^{14} 5d^{10} 6p^3$

For atoms with many electrons, it becomes tedious to write out the complete order of filling. It is customary in these cases to explicitly indicate only the electrons beyond the previous noble gas. The presence of the inner electrons is indicated by writing the noble gas symbol enclosed in brackets.

Using this shorthand system, the electron configurations for Ca, Fe, and Bi become

Ca $[Ar]4s^2$

Fe $[Ar]4s^23d^6$

Bi $[Xe]6s^24f^{14}5d^{10}6p^3$

PERIODIC PROPERTIES ELECTRON CONFIGURATIONS FROM THE PERIODIC TABLE

The periodic table was formulated long before the advent of quantum theory, but the ordering of elements was understood only on an experimental basis. Quantum Theory provides an explanation for the appearance of the periodic table in terms.

Of the order of filling of orbitals. In fact, this order of filling may be read directly from the table. We recognize correlations between its appearance and the order of filling given above:

- A new row (period) of the table begins each time occupation of a new main shell begins. The period number is the value of n for the new main shell.
- The block of elements in the two left-most columns of the table are those in which the ns subshell is being filled. This is the s-block.
- The block of elements in the 6 right-most columns are filling the np subshell. These constitute the p block.
- The block consisting of 3 rows of elements, each 10 blocks long, are filling the (n–1) d subshell. This is called the d-block.
- The remaining block of 2 rows, 14 elements long, are filling the (n–2) f subshell. This is called the f block.

We now show by example how to "read" the electron configuration of an element from the periodic table.

Example. Determine the electron configuration of thallium, Tl, from its position in the periodic table.

Solution. We read the configuration by scanning the rows (periods) one by one, top to bottom, until we reach thallium.

Period: $1s^2$ (there are only 2 elements in this row, H and He, corresponding to filling the n = 1 shell).

Period: $2s^2\ 2p^6$ (there are 8 elements in this row, corresponding to filling the 2s and 2p subshells of the n = 2 main shell).

Period: $3s^2\ 3p^6$ (again, scanning left to right we find 8 elements).

Period: $4s^2\ 3d^{10}\ 4p^6$ (Here we find 18 elements as we scan left to right. We encounter for the first time a row of d block elements in this period. The first 2 elements in the period give us $4s^2$, the d-block row gives us $3d^{10}$, the p-block row gives us $4p^6$.)

Period: $5s^2\ 4d^{10}\ 5p^6$ (Similar to the 4th period).Period 6: $6s^2\ 4f^{14}\ 5d^{10}\ 6p^1$ (Scanning left to right, we encounter 2 s block, 14 f block, and 10 d block elements. Thallium is the first p block element in this row, hence $6p^1$.)

Now string these together in order to obtain the complete configuration. To obtain the configuration in shorthand notation, we need read only from the previous noble gas. Thus

$$[Xe]\ 6s^2\ 4f^{14}\ 5d^{10}\ 6p^1$$

You are certainly free to memorize the order of filling if you want to. However, with practice it is much quicker and easier to read the configuration from the periodic table.

There remains one matter with which we must deal before moving on. The electron configuration for nitrogen is given below.

$$N\ 1s^2\ 2s^2\ 2p^3$$

Because there are three 2p orbitals, it is not clear from the configuration how the three electrons are to be distributed in them. There are several possiblities, a few of which are shown here. It is found experimentally that the last of these arrangements, with one electron in each of the 2p orbitals and all electron spins parallel, is preferred. Generally, when several electrons occupy a set of equal-energy (degenerate) orbitals, the most stable arrangement is the one with the maximum number of unpaired electrons.

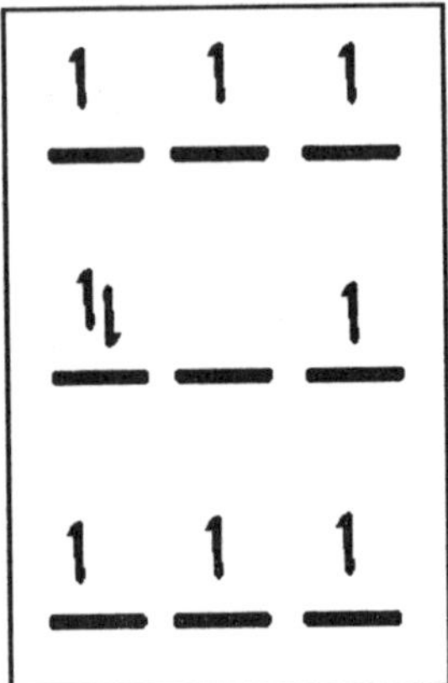

Fig. Electron Arrangements for Nitrogen

This statement is known as Hund's Rule. In accordance with Hund's Rule, the valence orbital occupations of carbon, oxygen, and iron are shown here.

C 2s 2p

O 2s 2p

Fe 4s 3d

Fig. Configurations for Carbon, Oxygen, and Iron.

Electron Shielding and Effective Nuclear Charge

We now return to an idea that we earlier stated as fact, without attempt at explanation. That is, that in an atom with several electrons, the energy ordering of subshells within a main shell is s < p (< d (< f)).

The core electrons (in 1s, 2s, and 2p) are shown as a sphere that is inside the distance of maximum probability for the n = 3 shell. The sodium atom has a single electron in the n = 3 main shell, which must occupy one of the three subshells in Figure 2-11. Experimentally, it is found to occupy the 3s subshell, which we conclude must be more stable (lower energy) than the 3p and 3d subshells. Why? We can make a few observations about these plots that will hopefully tell us.

- The maxima in the 3s, 3p, and 3d radial plots all occur at approximately the same radius. However, the probability of finding the electron at distances closer than this most probable distance is greatest for the 3s subshell and least for the 3d subshell. We make this judgment based on the magnitude of the probability at distances less than the curve maximum. The electron in the n = 3 shell spends part of its time inside the electrons in lower lying subshells (1s, 2s, and 2p) because the radial curves in Figure above are non zero inside the sphere representing the core electrons. We say that the n = 3 electron penetrates the core electron cloud.

The parentheses remind us that these subshells do not occur in shells with n < 3 or 4. We will try to develop an understanding of why this is true in terms of the radial plots for the 3s, 3p, and 3d subshells of the sodium atom shown qualitatively in Figure.

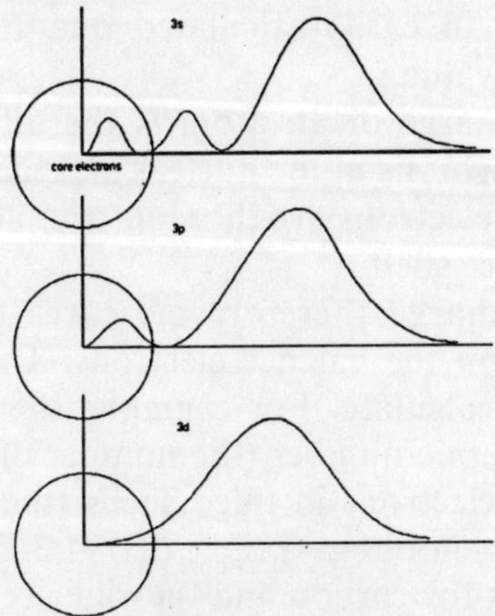

Fig. Core Penetration

- Generally, the greater the amount of penetration, the

more on average the electron feels the positive attraction of the nucleus, and the lower its potential energy. The amount of penetration is greatest for the 3s subshell and least for the 3d. The electron therefore occupies 3s.

The penetration of the core electron cloud by the 3s electron is of course incomplete, so the 3s electron does not feel the full effect of the 11 positive charges in the nucleus. The core electrons shield the 3s electron to some extent from the nucleus by coming between them.

The net positive charge felt by the 3s electron may be called the effective nuclear charge, Z_{eff}. It is the difference between the number of protons in the nucleus (the atomic number, Z) and the average number of electrons between the 3s electron and the nucleus, which we symbolize S:

$$: Z_{eff} = Z–S$$

S, the so-called shielding factor, is related to but always less than the number of shielding electrons. It is less than the number of shielding electrons because the shielded electron penetrates the core to some extent. Calculation of the value of S for a particular atom is somewhat complex; consequently, we will not attempt it.

Instead, we will discuss a quantity called the core charge, Z_{core}, which is an easily-calculated quantity that roughly parallels Z_{eff} in value.

The core charge of an atom is the difference between the number of protons in the nucleus and the number of core electrons. Core electrons are those in completely filled shells below the valence shell.

The core charge therefore measures the net positive charge pulling on the valence electrons. The core charge is very simple to calculate. For example, the core charge for lithium is its atomic number (the number of protons) minus the number of electrons in filled shells (the atomic number of the preceding noble gas): Z_{core}(Li) = 3–2 = 1. Similarly, the core charges for carbon and fluorine are 6-2 = 4 and 9-2 = 7, respectively. Core charges for the period 3 elements are worked out below:

Element	*Z*	*Number of core electrons*	Z_{core}
Na	11	10	1
Mg	12	10	2
Al	13	10	3
Si	14	10	4
P	15	10	5
S	16	10	6
Cl	17	10	7
Ar	18	10	8

Note that the core charge for an element is equal to the number of valence electrons (which is the last digit of the group number). Two trends in Z_{core} are of importance to us. First, Z_{core} increases as we proceed left to right across a row of the periodic table, because the number of protons in the nucleus increases while the number of core electrons stays the same.

Second, Z_{core} remains constant down a family of the periodic table, because the elements in a family all have the same number of valence electrons. These simple trends are the basis for the experimentally observed variations in atomic size, ionization energy, and other atomic properties discussed below.

Atomic Size

The molar volume for an element, in units of mL/mole can be calculated readily by dividing the atomic weight in g/mole by the density of a condensed phase (solid or liquid) of the element in g/mL. When molar volume is plotted as a function of atomic number, Figure results.

Examination of the plot shows two clear trends. First, molar volume tends to decrease from left to right across a period. Second, molar volume increases top to bottom down a family. If we make the reasonable assumption that there is a relationship between the molar volume and the size of an atom of the element, then the second trend is expected, but the first is counter-intuitive. Most of us would probably predict that atoms should get larger as the number of electrons increases across a period.

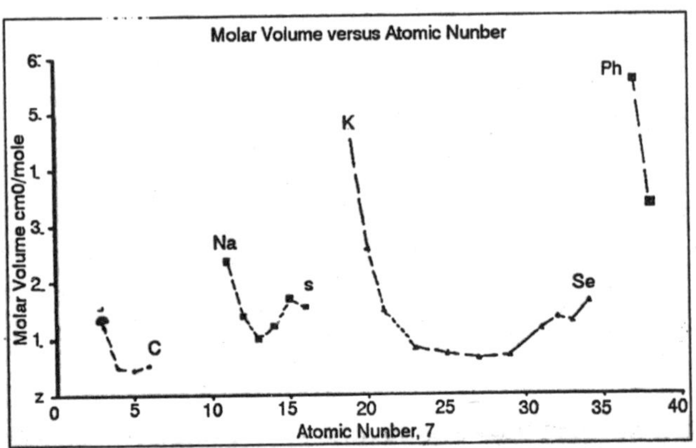

This is quite readily understood in terms of the trend in Z_{core} discussed above. Across a row, the added electrons enter the same main shell while experiencing an ever increasing pull from the nucleus.

The result is a steady shrinkage of the electron cloud. The elements of a family (column) of the periodic table have the same number of electrons in the shell of largest n. For example, the elements of group 2 all have 2 electrons in the outermost main shell. But they apparently do not. The results of x-ray diffraction experiments have given us fairly reliable values for the radii of atoms of many elements. Some representative values are presented in Table. It is clear from the data that atomic radii do indeed decrease left to right across a row of the periodic table.

Table. Atomic Radii, pm (Values are Van der Waals radii)

H	132						
Li180	Be---	B---	C168	N155	O150	F155	Ne 160
Na230	Mg170	Al---	Si210	P185	S180	Cl180	Ar190
K 280	Ca---				Se190	Br190	Kr 200
							Xe 220

Thus the core charge is the same for all members of a family. The vertical trend in atomic size therefore reflects the increasing distance from the nucleus that accompanies an increase in the value of the principle quantum number, n. Before ending this discussion of periodic trends in atomic size, it is useful to make a rough calculation of the size of an atom from the experimentally measured atomic volume. Mercury, a liquid metal, has a density of 13.6 g/mL and atomic weight 200.6 g/mole. The molar volume of mercury is thus

$$(200.6 \text{ g/mole})/(13.6 \text{ g/mL}) = 14.8 \text{ mL/mole}$$

We divide by Avogadro's number to convert this to an atomic volume:

$$(14.8 \text{ mL/mole})/(6.02\times10^{23} \text{ atoms/mole}) = 24.5\times10^{-24} \text{ cm}^3$$

To approximate the atomic radius, we take the cube root of the atomic volume:

$$(24.5\times10^{-24})^{1/3} = 3\times10^{-8} \text{ cm}$$

This very simple calculation provides a very approximate experimental value for the size of the atom: 10^{-8} cm, or 100 pm.

Ionization Energy, I

The energy required to remove the least tightly held electron of an atom in the gas phase is called the first ionization energy, I_1, of the atom. The process of electron removal is shown for the Li atom in equation:

$$Li(g) \rightarrow Li^+(g) + e- \text{ Energy required} = I_1 > 0$$

The electron removed in this case is the single 2s valence electron. Figure shows a plot of the first ionization energy versus atomic number for the first 3 periods of the periodic table.

The following trends are recognizable from the plot.

- Ionization energy tends to increase left to right across a period. For example, look at the section of the curve corresponding to period 2 (Z = 3 through 10). However there are downward "jogs" at boron and at oxygen.
- Ionization energy smoothly decreases down a family, with no jogs.

The force exerted on the outer electron by the core charge that it feels is given by Coulombs Law:

$$F = -Z_{core}e^2/r^2$$

Here Z_{core} is the core charge felt by the electron, e is the charge of the outermost electron, and r is the average distance of this electron from the nucleus. It is clear from equation that the force, and therefore the required input energy, increases as Z_{core} increases. We have seen above that Z_{core} increases steadily from left to right across a period. Thus F and I_1 are expected to increase also.

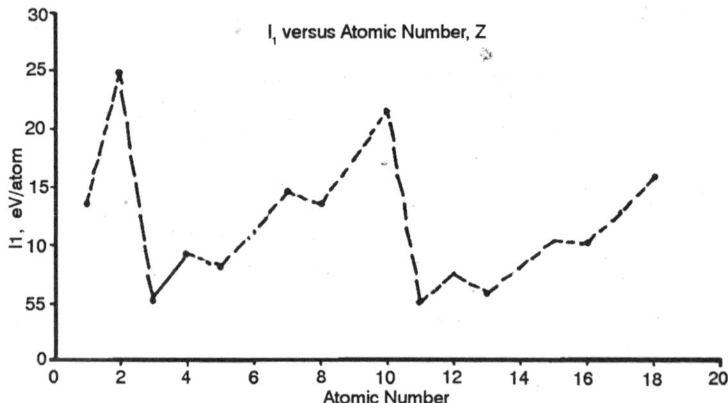

Fig. Ionization Potential versus Atomic Number

The overall increasing trend is readily understood. Similarly, the force decreases as r increases. Since we have seen that r increases with increasing n down a family, the decrease in I_1 down the family is also understandable in terms of 2-6-3. For the time being we will not concern ourselves with the "jogs" in the left-to-right trend in I_1. We will provide an explanation for them at the point that they become important.

It is of course possible to remove more than one electron from an atom. Sequential removal of the first 3 electrons from an atom, E, is represented in equations:

$E(g) \rightarrow E^+(g) + e$ Energy required = I_1

$E+(g) \rightarrow E^{2+}(g) + e$ Energy required = I_2

$E^{2+}(g) \rightarrow E^{3+}(g) + e$ Energy required = I_3

The energies required to remove the second and third electrons are called the second ionization energy, I_2 and the third ionization energy, I_3, respectively. No matter what particular atom E represents, it is invariably true that $I_3 > I_2 > I_1 > 0$. Of course, it requires the input of energy to remove the outermost electron of the atom because the electron must be pulled away from the effective positive nuclear charge that it feels. The second electron is therefore harder to remove because it must be pulled away from the positive ion, E^+. For the same reason, the third electron is even harder to remove than the second. The first 3 ionization energies for the elements through Ar are given in Table.

Table: Ionization Energies and Electron Affinities of the Elements, kJ/mole

Element	$EA_2(I_{-1})$	$EA_1(I_0)$	I_1	I_2	I_3
H	---	---	1312		
He	---	---	2372	5250	
Li	---	59.6	520	7298	11815
Be	---	-241	899	1757	14848
B	---	26.7	801	2427	3660
C	---	121.85	1086	2353	4621
N	-801	<0	1402	2856	4578
O	-780	140.98	1314	3388	5300
F	---	328.0	1681	3374	6051
Ne	---	-28.9	2081	3952	6122
Na	---	52.87	496	4562	6912
Mg	---	-231	738	1451	7732
Al	---	42.55	578	1817	2745
Si	---	133.6	786	1577	3231
P	-463	72.0	1012	1903	2912
S	-590	200.4	1000	2251	3361
Cl	---	349	1251	2297	3822
Ar	---	-34.7	1520	2666	3931

When the first electron is removed, the remaining electrons are pulled in more tightly to the nucleus because there is now a net positive charge of 1 unit on the atom.

Example: The ionization energy of hydrogen is 1312 kJ/mole. This is roughly the amount of energy that you would spend in curling a 220-lb (100 kg) barbell 1000 times. How many times would you have to curl the barbell in order to produce 1 mole of Mg^{2+} ions from Mg atoms? Solution. The sum of I_1 and I_2 for Mg is 2189 kJ/mole. This is 1.67 times larger than I for hydrogen. 1700 curls will do it.

Study of the table provides an explanation for a

statement presented. Elements in Groups 1, 2, and 13 of the periodic table form cations with positive charge equal to the last digit of the group number.

Another way to say this is that in forming compounds, these elements lose all of the electrons in the outer main shell, but do not lose any of the electrons in lower shells.

Successive values of the ionization energies for these elements show why. For Mg, I_2 is larger than I_1, for reasons discussed above, but by a factor of only 2. Loss of two electrons depletes the outer main shell of the magnesium atom.

If further electrons are to be removed, they must be taken from the next lower (n = 2) shell, which is substantially closer to the nucleus than the n = 3 shell.

Consistent with this, I_3 is more than 5 times larger than I_2! The cost in energy to remove the third electron is too high. Mg therefore stops at the Mg^{2+} cation. The sodium atom has only one electron in the outer shell. Its removal requires the energy, I_1, which is only 496 kJ/mole. However, removal of a second electron from Na+ requires disruption of the n = 2 shell. Roughly 10 times more energy is required to accomplish this than is required to remove the first. Again, the cost is too high, and sodium stops at Na+.

Electron Affinity, EA

The electron affinity is closely related to ionization energy. It provides a measure of the tendency of an atom to gain, rather than lose an electron. The first electron affinity, EA_1, is defined as the energy required to remove an electron from the anion, E-, forming the neutral atom E:

$$E\text{-}(g) \rightarrow E(g) + e \quad \text{Energy required} = EA_1$$

This equation is written analogously to equations; that is, it shows loss of an electron by a species, forming a related species having one more positive (or one less negative) charge. Thus the first electron affinity is sometimes called the zeroth ionization energy, I_0 (because a species of zero charge is formed). The second electron affinity (also I_{-1}) is defined analogously:

$E^{2-}(g) \rightarrow E\text{-}(g) + e$ Energy required = EA_2 or I_{-1}

As explained above, successive removal of electrons becomes progressively more and more difficult because the electron must be pulled away from a centre of increasing positive charge. This is no less true of species that are initially negatively charged. Thus I_{-1} and I_0 fit in the expected way into the series of ionization energies:

$$I_3 > I_2 > I_1 > I_0 > I_{-1}$$

In contrast to ionization energies, however, electron affinities are not always positive. That is, for some species, E-, the electron comes off spontaneously. No energy need be put in to cause it to happen.

Instead, energy is produced as the electron comes off. EA_1 is positive for some elements toward the right of the periodic table, but is in fact negative for many elements. EA_2 is negative for all elements.

Just as the trend in values of ionization energies serves to rationalize the observed positive charges of cations from groups 1,2, and 13, electron affinities rationalize the complementary rule: that elements from groups 15-17 tend to form anions with a number of negative charges equal to (8–group number).

Another way to say this is that these elements can add electrons to the point of filling the current main shell, despite the fact that EA is assuredly negative for any electrons past the first one added; but they will not add more electrons than this, because these would have to occupy a new main shell, further from the nucleus, where the added electrons feel a greatly diminished attraction from the nucleus.

Supplement. Experimental Evidence for the Existence of Shells—Photoelectron Spectroscopy. Spectroscopy is the study of the interaction of light and matter. Our understanding of atoms and molecules is largely attributable to spectroscopy.

In this section, we briefly discuss Photoelectron Spectroscopy, PES, in which light is used to eject electrons from atoms (or molecules). By subtracting the kinetic energies of the ejected electrons from the energy of light used to eject them, the ionization energies of the electrons can be calculated:

Energy of photon = ionization energy of electron + kinetic energy of electronhn = I + $KE_{electron}$

By varying the frequency of light used, electrons from various shells of the atom can be removed. This gives a measure of the energies levels of the shells. This idea is illustrated schematically in Figure.

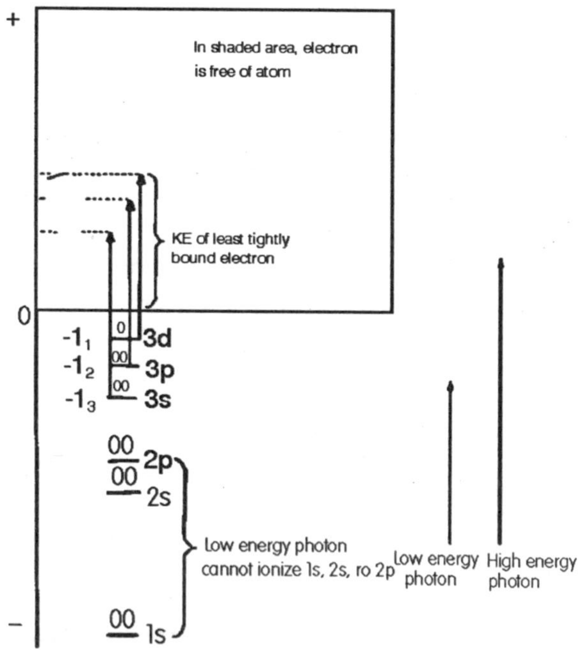

Fig. Photoelectron Spectroscopy

Table shows ionization energies obtained by photoelectron spectroscopy for the first 10 elements of the periodic table, H through Ne. Let's see what we can make of these data.

First, we see that H has only one ionization energy; we would expect this, because it has only one electron. Similarly, He has only one ionization energy, even though it has two electrons, but the intensity of the photoelectron signal is twice that for hydrogen.

This indicates that there are twice as many electrons susceptible to ejection by a photon of light. A single ionization energy is understandable if the two electrons are just alike; that is, if they occupy the same subshell. This is consistent with our understanding of helium, for which we write the electron configuration $1s^2$. We also see that the single ionization energy for He is larger than that for hydrogen. This is perfectly consistent with its greater core charge, Z_{core}.

For Li, there are two distinct ionization energies with intensities in the ratio 1 to 2. The first ionization energy is of relatively small magnitude, corresponding to an electron that is easily removed. The second ionization energy is quite substantial in magnitude, corresponding to two electrons that are much more difficult to remove. This is in accord with the quantum picture, in which there is a single electron in the outer n=2 shell, and two electrons in the inner filled n=1 shell. The PES for Be gives similar information.

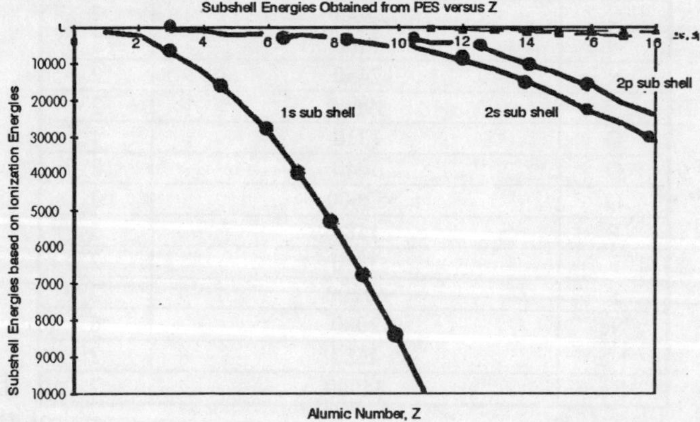

However, for boron, three types of electrons are observed, in relative numbers 1, 2, and 2. There is a single electron that is relatively easily removed.

There are then two electrons that are also relatively easily removed (about the same as the single electron of hydrogen), though more difficult than the first. Finally, there is a pair of electrons that is very tightly held by the nucleus of the boron atom. These results are consistent with a picture in which the the three easily ionized electrons are in two different subshells of the n = 2 main shell. The remaining two electrons are then deeply buried in the core shell with n = 1. Proceeding from B to.

Table: Results of Photoelectron Spectroscopy for H through Ne

Element	*Ionization Energy of Electron, kJ/mole*	*Intensity*
H	1310	1
He	2370	2
Li	520	1
Li	6260	2
Be	900	2
Be	11500	2
B	800	1
B	1360	2
B	19300	2
C	1090	2
C	1720	2
C	28600	2
N	1400	3
N	2450	2
N	39600	2
O	1310	4
O	3040	2
O	52600	2
F	1680	5
F	3880	2
F	67200	2
Ne	2080	6
Ne	4680	2
Ne	84000	2

Ne shows up to six electrons entering the second subshell of the n = 2 main shell. Figure shows a plot of the energy levels of the ionized electrons against the atomic number of the first 18 elements. (The energies of the levels are the negatives of the ionization energies, which measure the amount of energy required to separate the electron from the nucleus. Recall that the energy of the electron is considered to be zero when it is completely separated from the atom;

thus the energy of the electron in the atom before its removal must be negative.)

The energy of a particular subshell as a function of atomic number is indicated by connecting the appropriate points with a curve. Several things are evident from the plot:

- The shell, subshell structure of the atom is obvious;
- The energy of a particular subshell (e.g., 1s) decreases very dramatically as Z increases;
- The gap in energy between main shells increases dramatically as Z increases;
- The gap in energy between the subshells of a particular main shell increases substantially as Z increases across a row of the periodic table.
- The ionization energies of electrons in the valence shell fall within a fairly narrow range (500-5000 kJ/mole, judging from Figure).

  This range is a measure of the energies involved in chemical reactions.

Chapter 3

Structures of Simple Molecules

LEWIS STRUCTURES OF MOLECULES

An atom of an element has a characteristic number of electrons, equal to its atomic number Z. These electrons are arranged in shells about the nucleus, some being close to the nucleus, others far away. It is the electrons in the outermost shell, furthest from the nucleus, and therefore least tightly held, that are involved in chemical bonding. These are the valence electrons.

The electrons close to the nucleus, called inner or core electrons, are much too tightly held to be lost to or shared with another atom. As we have seen, the number of valence electrons that an atom has is equal to the last digit of its group number.

Thus Na, in group 1, has one valence electron; Al in group 13 has three valence electrons; Br in group 17 has seven valence electrons; and Fe in group 8 has eight valence electrons. The Lewis symbol for an element shows the valence electrons as dots arranged about the symbol for the element, for Na, Al, and Br. Dots are placed as if the symbol of the element were surrounded by an invisible square.

The first dot can be placed on any of the 4 sides, but the second, third, and fourth dots are then placed singly on the remaining 3 sides of the square; electrons are not placed in pairs (i.e., 2 dots on the same side of the square) until 4 have been placed singly. Dots that singly occupy the side of a square are referred to as unpaired electrons.

G.N. Lewis developed this approach to representing the

valence electrons of an element before the quantum view of the atom was complete. It is of interest from our perspective, though, to examine the correspondence between the Lewis symbol for an element and its valence electron configuration. In particular, we might expect the number of unpaired electrons in the Lewis symbol to correspond with the number of electrons that singly occupy valence orbitals of the atom. The Lewis symbols and valence orbital occupations of the atoms of period 2.

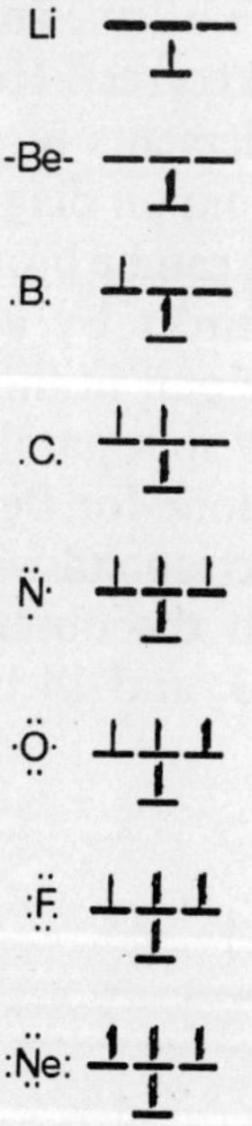

Li

·Be·

.B.

.C.

N

O

F

Ne

Fig. Valence Electrons for Period2

What we see is interesting and a bit puzzling. First, most of the atoms show the intuitively expected correspondence. Thus Li, N, O, F, and Ne exhibit the same number of unpaired electrons in the Lewis symbol and in the quantum orbital occupation diagram.

Further, the numbers of unpaired electrons for these elements correspond with their usual valences; i.e., with the number of bonds that they form.

However, for Be, B, and C in groups 2, 13, and 14, the

numbers of unpaired electrons in the Lewis symbols do not correspond with the orbital occupation diagrams. Thus the Lewis symbol for carbon leads us to expect 4 unpaired electrons; the orbital occupation diagram predicts 2. The observed valence for carbon is (almost) invariably 4, in accord with the Lewis symbol. Similarly, the valences of Be and B are as expected from the Lewis symbols, but in conflict with the orbital occupation diagrams.

Yet the success of the quantum theory of the atom has been so profound that we believe completely that the orbital occupation diagrams are correct. How are we to explain the bonding properties of elements in groups 2, 13, and 14 in terms of the orbital occupation diagrams?

The orbital diagrams can be brought into correspondence with the observed valences by a simple expedient: the downspin electron of the pair in the s orbital is placed in an empty p orbital, with its spin parallel to those of the other electrons. When this is done for Be, B, and C, the numbers of unpaired electrons correspond with the predictions of the Lewis symbol, and with the observed valences. Thus the elements of groups 2, 13, and 14 behave as if their orbital occupation.

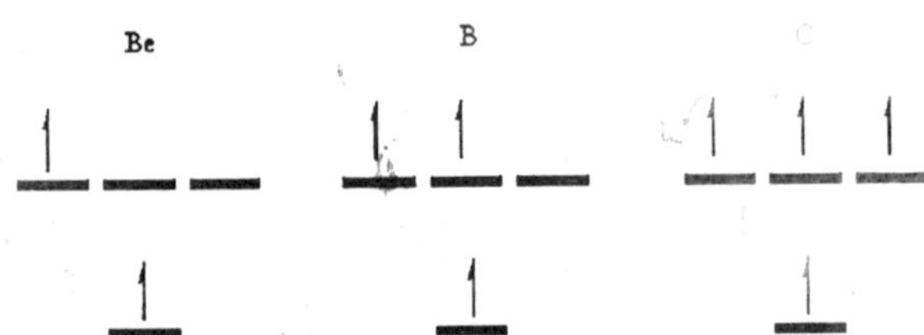

Fig. Apparent Configurations for Be, B, and C

The elements in group 18, called the Noble Gases, have 8 valence electrons, arranged in pairs on the 4 sides of the Lewis symbol square. The one exception is helium, He, which has 2 valence electrons. As their name indicates, these elements occur in nature as very unreactive gases; this implies that 8 valence electrons, called an octet, is in some sense a stable situation. As we saw that, 8 electrons is sufficient to fill the s and p orbitals of a shell.

When the atoms of the elements undergo chemical reactions, the frequent result is a filled set of valence s and p orbitals for each atom involved — i.e., each atom acquires the octet of valence electrons possessed by the noble gases. The sodium atom usually loses its single valence electron upon reaction, forming Na^+.

The sodium ion has the same number of electrons as the noble gas, Ne. The Al atom must lose 3 valence electrons to achieve a noble gas configuration, so it forms the Al^{3+} ion. In contrast, Br may much more readily gain a single electron to achieve the same number of electrons as Kr than lose 7 electrons. It therefore is found frequently in compounds as the Br- ion.

Elements with from 3 to 7 valence electrons often achieve an octet of electrons—a full set of valence s and p orbitals—by sharing, rather than gaining or losing, electrons. For example, the Br atom can achieve an octet if it can gain 1 electron by sharing. A second Br atom can supply this electron, and achieve the octet for itself in the process. Shared electrons contribute to the octets of both of the two sharing atoms.

The shared pair of electrons, located between the bromine atoms, is circled. Note that the single unpaired electron of one Br atom is paired with the single unpaired electron of the second Br atom. This pair is placed between them to indicate the sharing.

The shared electron pair constitutes the bond between the two atoms, because it is attracted to both nuclei and therefore holds them together. The bond is called a covalent bond, because it involves cooperation between the 2 atoms so that each may achieve an octet. It is customary and convenient to represent electron pairs with lines rather than dots,. Keep in mind that each line represents 2 electrons an electron pair.

Frequently, the number of covalent bonds that an atom forms is the same as the number of unpaired electrons in its Lewis symbol (we will see subsequently that many atoms can be represented by more than one Lewis symbol, giving them some flexibility in the number of covalent bonds formed). Therefore Br, and the other halogens, commonly form one covalent bond. Oxygen has two unpaired electrons and can achieve an octet by sharing both of these with another oxygen atom.

The two oxygen atoms then share a total of four electrons, each thereby achieving the octet. Two pairs of electrons shared between two atoms gives two bonds; the pair of bonds together is called a double bond, and is represented by a double line as in the structure for O_2. You should satisfy yourself that two nitrogen atoms can achieve an octet by sharing three pairs of electrons to form a triple bond.

Carbon, with 4 unpaired electrons, forms 4 covalent bonds. Similarly, sulfur forms 2, phosphorus 3, silicon 4, hydrogen 1, and so on, based on the dot. Hydrogen is an exception to the octet rule, because it can accommodate only 2 valence electrons. In acquiring 2, it achieves the electron configuration of the noble gas, helium.

Lewis Structures for Molecules

The realization that when elements react to form compounds the atoms frequently achieve stable electron arrangements octets was first made by G.N. Lewis, a great American chemist. In tribute to the contributions made by Lewis to the theory of chemical bonding, molecular structures in which valence electrons are represented by dots or lines are called Lewis structures. The Lewis approach to structure, called Lewis theory, is our most-used theory of molecular structure. It is simple, yet sophisticated enough to provide an understanding of molecular properties such as shape, bond length, bond angle, and bond strength.

Further, it allows explanation of chemical reactivity and mechanism in terms of structure for many chemical reactions, and suggests a general approach to acids and bases that is used widely. In due course, we will address all of these matters. In this section, we develop a systematic approach for obtaining the Lewis structures of molecules. Some Lewis structures are easily obtained from the dot pictures of the constituent atoms. Dot pictures for carbon and fluorine. Clearly carbon is able to form four bonds, and each fluorine is able to form one. Therefore, one C atom bonds to four F atoms.

The Lewis structure for CF_4, tetrafluoromethane. Similarly, the Lewis structures for ammonia and water are simply obtained. In these molecules, each atom forms a number of covalent bonds equal to the number of unpaired electrons in its dot picture, and each atom has an octet (a duet in the case of hydrogen, which gives it the configuration of He, with 2 valence electrons) in the final structure.

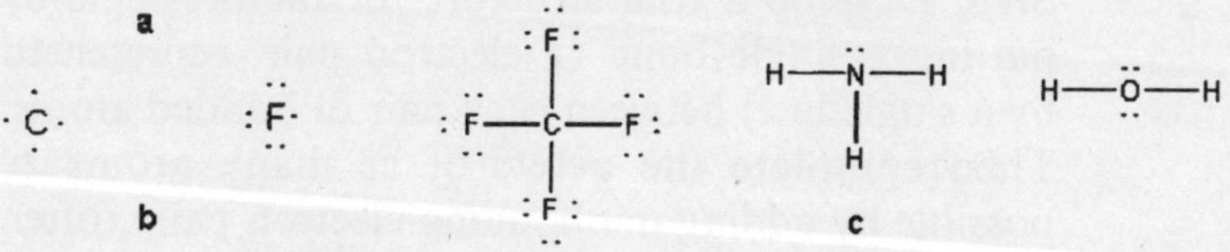

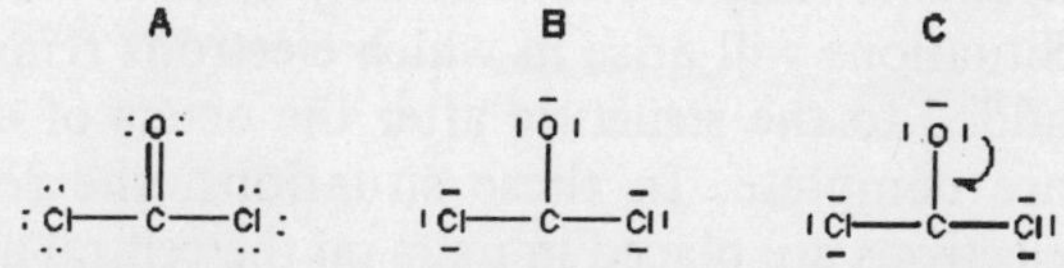

The Lewis structure is not always so readily obtained, however. For example, that for phosgene, $COCl_2$. Although inspection of this structure verifies that each atom has an octet and has formed the expected number of bonds, arriving at this structure from scratch requires some thought.

Note the presence of a double bond between the carbon and oxygen atoms. How are we to know when double bonds and triple bonds, collectively called multiple bonds, are appropriate? To address this question, we now present a systematic and straightforward approach to the writing of Lewis structures.

- *Step:* Arrange the atoms in the correct layout (i.e., bonding arrangement). The layout is experimentally determined, but can often be deduced from the formula for the molecule (or ion). For many molecules (but not all!), the first listed atom is in the centre, and the others are attached to it. This is the case for phosgene.

- *Step:* Count the number of valence electrons in the molecule by summing the electrons contributed by each atom, based on their group numbers. If the molecule is negatively charged, add the corresponding number of electrons. Subtract the appropriate number of electrons if the molecule is positively charged.
  For phosgene,
  Number of valence electrons = 6(O) + 4(C) + 2 × 7(Cl) = 24.
- *Step:* Develop a trial structure for the molecule by putting a single bond (1 electron pair, represented by a single line) between each pair of bonded atoms. Then complete the octets of as many atoms as possible by adding nonbonding electron pairs (often called lone pairs) to the atoms in turn until all of the available valence electrons (step 2) have been used. Situations will arise in which electrons remain to be added to the structure after the octets of all atoms are complete. In these situations, the remaining electrons are placed in pairs on the central atom.
- *Step:* Examine the trial structure. If all atoms have octets (or if the terminal atoms have octets while the central atom has more than 8 electrons), the trial structure is acceptable. If the trial structure shows atoms with fewer than eight electrons, it must be altered by using lone pairs to form multiple bonds until all atoms have octets.

The phosgene structure must clearly be amended because the carbon atom has only six electrons in the trial structure. A lone pair from the oxygen atom is moved in to form a second bond, preserving the octet on oxygen and at the same time establishing the octet for carbon.

The result is an acceptable Lewis structure for phosgene. It is important to state at this point that an acceptable structure obtained by direct application of the procedure above is not necessarily the optimal (best) structure.

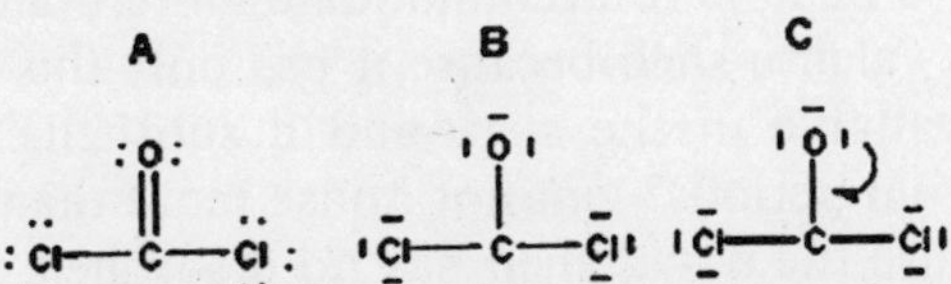

We emphasize that once the correct number of electrons is placed in the structure in step 3, no more may be added, and none may be removed; electrons can, however, be moved. (A lone pair of oxygen is preferred over one from chlorine for reasons dealing with the concept of formal charge.)

Exceeding the Octet

Although we have used the octet rule as the basis for development of a systematic approach to Lewis structures, only the atoms of period 2 of the periodic table obey it (almost) without exception. Any atom having valence shell principal quantum number ^{3} 3 may exceed the octet (violate the octet rule).

When the octet of electrons is exceeded, it is usually at the central atom. An atom can exceed the octet only if it has atomic orbitals available for housing the additional electrons. As we have stated, eight electrons is sufficient to fill the s and p subshells of the valence shell (outermost main shell) of the atom.

To accommodate more than eight electrons in the valence shell, an atom must have available orbitals in addition to the s and p. Atoms for which the valence shell has n ^{3} 3 have a valence d subshell comprised of five d-type atomic orbitals. When an atom exceeds the octet, it makes use of d orbitals to house the additional electrons.

The Se atom in SeF_4 uses a total of 5 valence orbitals (one s, three p, and one d) to accommodate the four bonding pairs and one lone pair of electrons that surround it. The sulfur atom in SF_6 uses 6 valence orbitals (one s, three p, and two d) to accommodate the six bond pairs.

And the Te atom in TeF_8^{2-} uses 8 valence orbitals (one s, three p, and four of the five d) to house eight electron pairs! Note that we would not expect an atom with valence shell n ³ 3 to be able to accommodate more than 9 electron pairs in its valence shell because it has only this number of orbitals available in the s, p, and d subshells. Similarly, elements from period 2 can not house more than 4 electron pairs, because the n = 2 shell has no d subshell. The atoms in period 2 are strictly limited by the octet rule.

Many atoms can be represented by more than one Lewis symbol. This is very closely connected with the availability of d orbitals in the valence shell, and we will now examine this matter before moving on. The sulfur atom is a good vehicle for illustration, because it shows a substantial amount of flexibility in the number of covalent bonds that it is able to form. The usual Lewis symbol for S, and the orbital occupation diagram with which it corresponds. This Lewis symbol allows us to rationalize the many 2-valent compounds that are known for sulfur, among them H_2S, SCl_2, and $(CH_3)_2S$. Formation of two covalent bonds is readily rationalized in terms of the two unpaired electrons in the Lewis symbol (and orbital occupation diagram). Thus in H_2S, each H atom pairs its single unpaired electron with one of the unpaired electrons of S, resulting in completed s and p subshells for S, and completed s subshells for H.

The other two-valent molecules are understood in similar ways. However, sulfur also forms a number of four-valent compounds (for example, SF_4); and six-valent compounds (for example, SF_6), which cannot be readily accounted for in terms of the simple Lewis symbol above. However, useful Lewis symbols for S can be obtained by unpairing electrons, one pair at a time. The resulting Lewis symbols then show four and six unpaired electrons.

Using these expanded Lewis symbols, we readily account for the flexivalent nature of the sulfur atom, and in fact for similar flexibility for most atoms with valence shell n ³ 3. The following generalization will prove useful to us in rationalizing many molecular structures:

A p-block atom usually forms compounds with a number of covalent bonds lying in the range between the number of unpaired electrons in its simplest Lewis symbol, and the total number of valence electrons that it has. Further, other common numbers of covalent bonds lie between these extremes, separated by increments of 2.

Applying this to S, we would predict that it should exhibit a minimum of 2 covalent bonds and a maximum of 6, with compounds containing 4 covalent bonds being common. Applying the guideline to chlorine enables us to predict a minimum of 1, a maximum of 7, with 3 and 5 covalent bonds being common. In fact, the compounds HCl (minimum), ClF_7 (maximum), ClF_3, and ClF_5 are well known.

A Formal Charge Refinement—Optimizing the Lewis Structure

We can combine our formal charge criteria for acceptable Lewis structures with the ability of many (central) atoms to exceed the octet to obtain improved structures for molecules in which the central atom is from period $n > 3$. By way of illustration, consider the structure of sulfuric acid. This structure results from application of the systematic approach to development of Lewis structures presented earlier. Although the sulfur and oxygen atoms obey the octet rule, the formal charges are a bit extreme.

Specifically, S has 2+ formal charge, and the two terminal oxygen atoms not bonded to hydrogen have formal charges of 1-. By using lone pairs on these two oxygens to form double bonds to S, the formal charges of all three atoms can be reduced to zero. This causes the sulfur atom to exceed the octet, but this is acceptable for an atom from period 3. The resonance form called the minimum formal charge structure, results. It is this structure that is usually drawn for sulfuric acid.

The minimum formal charge structure for any molecule can be obtained in a straightforward manner. Starting with the structure generated using the systematic procedure, assign formal charges to all atoms. If the central atom is not from period 2, introduce double bonds between terminal atoms of negative formal charge and the central atom until the central atom formal charge is zero. One double bond removes one unit of positive formal charge from the central atom, so to achieve zero formal charge requires a number of double bonds equal to the positive formal charge in the initial structure. Initial and minimum formal charge structures for H_4SiO_4, H_3PO_4, and $HClO_4$ are shown.

THE THREE-DIMENSIONAL STRUCTURES OF MOLECULES: VSEPR

Although Lewis structures give us a feel for the manner in which electrons are distributed and shared in molecules, they fail to convey a sense of molecular shape — the 3-dimensionality of molecules. Thus the Lewis structure for water makes it appear as though the molecule is linear; in fact, it is bent, like a V. Similarly, methane is not a square and planar molecule, as the Lewis structure implies, it is instead tetrahedral.

The experimental determination of the 3-dimensional shape of a molecule, whether it be a relatively simple molecule like water or a complex and large molecule like hemoglobin, the oxygen carrying protein of blood, involves a combination of spectroscopic methods and the methods of x-ray diffraction.

Based on the experimentally-determined structures of large numbers of molecules, it has been possible to develop an extremely simple theory that allows the prediction of 3-dimensional structure with a high degree of success. This theory is called the Valence Shell Electron Pair Repulsion (VSEPR) Theory. It is based upon a single very simple tenet: electron groups surrounding an atom in a molecule distribute themselves to minimize the repulsions between them. In other words, they get as far apart as is geometrically possible. We must establish three things in order to apply this idea. First, what constitutes an electron group? Second, what geometrical arrangement minimizes repulsions between a particular number of such groups? Third, once the groups are distributed, what is the molecular shape? An electron group is any of the following: a lone (nonbonding) pair of electrons;

a single bond; a double bond; a triple bond; a single electron. Counting the number of groups around an atom in a molecule is therefore quite easy.

Example. How many electron groups surround the central atom in each molecule? CH_4, H_2O, PF_5, $COCl_2$, NO_2.

Solution. A correct Lewis structure is needed before VSEPR can be applied. Lewis structures for PF_5 and NO_2, developed by the procedure. You should verify each structure to test your understanding of the procedure for obtaining Lewis structures. The number of electron groups around the central atom in each species can be counted:

H_2O	4
CH_4	4
PF_5	5
$COCl_2$	3 (the double bond counts as a single group!)
NO_2	3 (the double bond and single electron count as single groups!)

The geometrical arrangements that minimize repulsions among up to 6 groups are given in Table, with structural representations.

Table: Vsepr Electron Group Distributions

Number of groups	*Distribution*	*Example*
2	linear	BeH_2
3	trigonal planar	$COCl_2$
4	tetrahedral	CH_4
5	trigonal bipyramidal	PF_5
6	octahedral	SF_6

Trigonal planar means that the central atom and the 3 groups around it all lie in the same plane, and that the 3 groups are arrayed in triangular shape about the central atom. Trigonal bipyramidal means what it says: 2 pyramids, each with a triangular base, placed base-to-base. An octahedral distribution means that the 6 groups form an octahedron about the central atom. An octahedron is a regular figure with 6 vertices and 8 equilateral-triangular faces. You can think of it as a base-to-base joining of 2 square-based pyramids. In this view, it is somewhat similar in origin to the trigonal bipyramid. Please take some time to learn the shape-descriptive terminology, and to develop a mental image of each of the common group distributions. We see from these examples that two situations arise. In the first, all electron groups on the central atom are used in bonding other atoms. This is the case for CH_4, PF_5, and $COCl_2$. In the second, only some electron groups are shared with other atoms; the remainder are unshared.

This is the case for H_2O and NO_2. In all cases, the distribution of electron groups is as indicated. However, the molecular shape is a description of relative atom positions. When the number of attached atoms is less than the number of electron groups, descriptions of electron distribution and shape are different. First, consider the situation in which all groups are shared with other atoms.

In these cases, the descriptors for group distribution are also used for shape. Thus, methane has a tetrahedral shape (as well as a tetrahedral group distribution), phosphorus pentafluoride has a trigonal bipyramidal shape, and sulfur hexafluoride has an octahedral shape.

When there are nonbonding electrons on the central atom, the number of attached atoms is less than the number of electron groups. We have previously seen numerous examples: ammonia, water, sulfur dioxide, and selenium tetrafluoride.

The shapes of such molecules are described using words that are indicative of where the atoms are. Nonbonding electrons, although important in determining group distribution and therefore shape, are ignored in the geometrical description of shape.

Thus for example, the group distribution about the nitrogen atom in ammonia is tetrahedral, just as in methane; but the molecule is said to have a trigonal pyramidal shape. This descriptor tells us where the atoms — one N and three H — are relative to one another. Similarly, the group distribution in water is tetrahedral, just as in methane and ammonia; but since there are only 3 atoms, which describe a "V" in space, the water molecule is said to be bent. Further examples for electron groups numbering from two to six are in Table.

Table. Shapes of Molecules with Number of Attached Atoms < Number of Electron Groups

# Groups	*# Atoms*	*Distribution*	*Shape*	*Example*
2	2	Linear	Linear	CO_2
2	1	Linear	Linear by Necessity	CO
3	3	Trigonal planar	Trigonal Planar	BF_3
3	2	Trigonal planar	Bent	$SnCl_2$
4	4	Tetrahedral	tetrahedral	CH_4
4	3	Tetrahedral	trigonal pyramidal	NH_3
4	2	Tetrahedral	bent	H_2O
5	5	Trigonal bipyramid	trigonal bipyramid	PF_5
5	4	Trigonal bipyramid	teeter-totter	SF_4
5	3	Trigonal bipyramid	T-shaped	BrF_3
6	6	Octahedral	octahedral	SF_6
6	5	Octahedral	square pyramidal	BrF_5
6	4	Octahedral	square planar	XeF_4

The three-dimensional structures of ammonia, sulfur tetrafluoride, bromine trifluoride, bromine pentafluoride, and xenon tetrafluoride.

The molecular structures are drawn using the so-called wedge-hash system. In this system, as many as possible of the bonds between central and terminal atoms are placed in the plane of the paper, and are represented as solid lines.

Bonds that must by restraints of geometry protrude out of this plane are represented using wedges. The wedge shape represents the bond as getting thicker, or fatter, as it approaches us. Finally, bonds that retreat behind the plane of the paper are represented using dashed or dotted lines. You should become skilled at using this notation to represent the three-dimensional structures of molecules.

The shape of a molecule is referred to as its stereochemistry. Thus methane, CH_4, is said to have tetrahedral stereochemistry about the carbon atom. Similarly, NH_3 has trigonal pyramidal stereochemistry, and SF_6 has octahedral stereochemistry.

Example. Describe and draw the shapes (stereochemistries) of the molecules in Example.

Solution. Electron group distributions and shapes follow.

Molecule	*Group Distribution*	*Shape*
H_2O	Tetrahedral	Bent
CH_4	Tetrahedral	Tetrahedral
PF_5	Trigonal Bipyramidal	Trigonal Bipyramidal
$COCl_2$	Trigonal Planar	Trigonal Planar
NO_2	Trigonal planar	Bent

Bond Lengths and Bond Angles in Molecules. In this section, we continue our discussion of the structures of molecules by saying something about the lengths of chemical bonds, and the angles that bonds make with each other about a central atom. Lengths are relatively easy to understand, so we discuss these first, in terms of a couple of simple ideas.

- Bonds tend to become longer as atoms become larger. Thus the bond between a carbon atom and a (very small) hydrogen atom is shorter than that between two carbon atoms, or between a carbon atom and an oxygen atom. The bond between two carbon atoms is shorter than that between a carbon and a chlorine atom. The C-H, C-C, and C-Cl single bond lengths are given below. These values are averages, of course, because the precise length of a bond depends to some extent on the other atoms in the molecule:

  C – H 109 pm
  C – C 154
  C – Cl 177

- Multiple bonds between two atoms are shorter (and stronger) than the single bond between the same two atoms. This is entirely reasonable; two atoms sharing four rather than two electrons between them will be bonded more tightly together, leading to a shorter distance and more strength. If we define the bond order between two atoms as the number of bonds between them, then distance decreases and strength increases as bond order increases. For example, consider the lengths of the single (bond order 1), double (bond order 2) and triple (bond order 3) bonds between two nitrogen atoms:

N – N 145 pm
N =N 125
N (3)N 110

Bond angles are also extremely important in three-dimensional structure, and it is important to be able to make a simple qualitative prediction of their magnitudes. Regularly-shaped molecules have the bond angles expected on the basis of the associated geometrical figure. Thus BF_3, a perfect equilateral triangle (trigonal planar), has angles of exactly 120° between adjacent B-F bonds. Methane, CH_4, is a perfect tetrahedron, with angles of 109.5° between the C-H bonds.

Phosphorus pentafluoride is regularly-shaped, but differs from BF_3 and CH_4 in that the P-F bonds are of two different types. The three fluorines that define the equilateral triangle where the two pyramids are joined are separated from one another by angles of 120°, just as are the bonds in BF_3. These are said to occupy equatorial positions. The remaining two fluorines, one above and one below the equilateral triangle, lie along a straight line that also passes through the phosphorus atom. These fluorines are said to be axial. The angle between an axial and an equatorial fluorine is 90°, and that between the two axial fluorines, 180°.

Bond Angle Refinements

The VSEPR Theory allows us to easily predict the overall shape of a simple molecule based on the Lewis structure, and bond angles for those highly symmetrical molecules in which all central atom electron groups are shared with terminal atoms, and in which all terminal atoms are the same. With the addition of just a few more simple ideas, the theory can be extended to allow us to say something about bond angles in less symmetrical situations as well.

We begin with some experimental facts. The electron group distributions in methane, ammonia, and water are tetrahedral. However, whereas the bond angle in methane is the expected 109.5°, that in ammonia is 107°, and that in water is still smaller at 104°. Why? The nonbonding electron pair that occupies one of the four positions around the nitrogen atom in ammonia belongs entirely to the nitrogen atom; it is not shared with any other atom.

It is therefore closer to nitrogen than are the 3 single bond pairs, and requires more room. The result is that the three bond electron pairs are repelled more by the lone pair than by each other, and are squeezed closer togther than in methane; the bond angle is reduced.

The effect is even greater in water, where two of the four groups are nonbonding pairs. Similarly, it is found experimentally that the angle between the two C-F bonds in COF_2 (similar to phosgene, $COCl_2$) is 108°, noticeably less than the ideal value of 120°. This is attributed to the double bond between C and O, which involves four electrons rather than two, and therefore takes up more space.

Finally, the O-N-O angle in NO_2 is 132°, somewhat larger than the ideal value of 120°. As might be expected, the single electron group on the central N atom exerts less repulsion than would a nonbonding pair, and less even than a bond pair of electrons. As a consequence, the angle between bond pairs opens up.

Based on these observations, the following generalizations may be made:

- Repulsion between lone pairs and bond pairs falls in the following order: LP-LP > LP-BP > BP-BP (LP stands for lone pair, BP for bond pair).
- Repulsion between multiple bonds and single bonds falls in the following order: MB-MB > MB-BP> BP-BP.
- Repulsion between a bond pair and a single electron is less than that between two bond pairs: BP-BP> BP-SE (SE stands for single electron).

With these ideas in hand, we now address the strange structures for SF_4 and BrF_3 in Table. There are 5 electron groups around the central atom in both species, so the group distribution is trigonal bipyramidal. In both cases, lone pairs, which exert the largest repulsions on other electron pairs, occupy equatorial rather than axial sites because by so doing they minimize 90° interactions with other electron pairs.

A lone pair in an axial position would have three such interactions (with the equatorial pairs); in an equatorial position, it has only two 90° interactions (with the axial pairs). Lone electron pairs occupy equatorial sites in the trigonal bipyramid. Similar arguments apply for square planar XeF_4, in which the two lone pairs are as far apart as possible at 180°.

Example. Say as much as possible about the bond angles and lengths in the following species: N_2O, SO_4^{2-}, SO_3^{2-}, O_3 (ozone), ClO_2.

Solution. The approach is to obtain a Lewis structure, considering all reasonable resonance forms; to determine the number of electron groups around the central atom; and to determine shape. Bond angles are assessed by considering relative strengths of repulsions between electron groups on the central atom. Finally, bond lengths are qualitatively assessed for each bonded pair of atoms by "averaging" the resonance forms.

Lewis structures are shown in. Based on these structures, the bond angles are developed as follows:

Molecule:	N_2O	SO_4^{2-}	SO_3^{2-}	O_3	ClO_2
# electron groups:	2	4	4	3	4
group distribution	linear	tetrahedral	tetrahedral	trigonal planar	tetrahedral
stereochemistry	linear	tetrahedral	trigonal pyramid	bent	bent
bond angles	180	109	< 109	<120	>109

It is possible to make a number of statements about bond lengths. The NN bond in N_2O is somewhere between double and triple, based on an average over the two acceptable resonance forms. It should have length intermediate between these extremes. Similarly, the NO bond length should be intermediate between that for the N-O and N=O bonds. For SO_4^{2-}, bond lengths should all be the same, and should be somewhat shorter than the S-O single bond length. Since S is a larger atom then O, the SO bond length should be longer than the O-O single bond.

For SO_3^{2-}, bond lengths should again be all the same, and should be slightly shorter than single bond lengths. They should be similar to the bond lengths in SO_4^{2-}.

For O_3 there are two equivalent (identical) resonance forms. Averaging over them, we conclude that each OO bond should be intermediate in length between the OO single bond (as in hydrogen peroxide) and double bond (as in O_2).

Finally, for ClO_2, the two ClO bonds should have the same length, appropriate to that of a double bond.

It may have occurred to you that there is a discrepancy between the experimentally observed bond angles in molecules (109, 120, and 180°) and the spatial distribution of the valence s and p orbitals. It is reasonable to assume that the valence orbitals of an atom play a role in the formation of bonds to other atoms; however, at this point we are unable to explain why orbitals directed at 90° to one another form bonds that are separated by 109, 120, or 180 degrees.

We will rely heavily on the principles of molecular stereochemistry in work to come. Please study and practice using these principles until you are comfortable with them.

CHARGE DISTRIBUTION IN MOLECULES ELECTRONEGATIVITY

The electronegativity of an atom was defined by Pauling as "the ability of the atom in a molecule to attract electrons to itself." Unlike the molar volume and ionization energy, the electronegativity is not a directly measurable physical property of an atom.

Rather, it is defined in a somewhat vague and qualitative way. Several ways of assigning a numerical value to the electronegativity of an atom have been developed. The most commonly used scale of electronegativity is that of Pauling, based on the experimentally measured values of bond energies. We will not concern ourselves with the details of how the scale was established. We will instead present the scale, make some general statements about the variation of electronegativity with position in the periodic table, and then proceed to use the electronegativity where appropriate. Suffice it to say at this point that the electronegativity of an atom, symbolized c, is roughly proportional to the first ionization energy of the atom. In other words, the more difficult it is to remove the outermost electron from an atom, the more likely it is that the atom will attract additional electrons to itself.

The Pauling electronegativities of the elements are presented. The following trends can be extracted:

- c increases left-to-right across a period. This is just what we expect given the parallel between c and I_1. However, there are no "jogs" in the trend, as there are for I_1.
- c decreases steadily down a family, again paralleling the trend in I_1.
- There is a rough equality of the electronegativities of an element and the element one column to the right and one row below it. Thus the electronegativities of Cl and O are 3.2 and 3.4, respectively. Those of N and S are also quite similar. This results because the increase in c due to the rightward move and the decrease in c due to the downward move roughly cancel.

Electronegativity is the primary concept on which we base our understanding of electron distributions among atoms in a molecule.

Bond Dipoles and Dipole Moment

The distribution of electrons in a molecule determines its shape. One frequent consequence of shape is that the centers of positive and negative charge in the molecule are displaced from one another, so that the molecule has a positive end and a negative end. Such molecules are said to possess total molecular dipole moments. It is important to be able to assess whether or not a molecule is polar; that is, whether it has a total molecular dipole moment. In this section, we will take some time to discuss the origins of total molecular dipole moments so that you will be equipped to do this assessment.

An arrangement in which a positive charge q+ and an equal negative charge q- are separated by a fixed distance, r, is called an electric dipole. The electric dipole is a vector quantity that is by convention taken to point from the positive to the negative end of the dipole. The magnitude (length) of the vector is the product of the charge, q+, and the distance between the positive and negative centers:

$$: m_d = qr$$

The electric dipole has units of charge times distance. For a dipole consisting of a charge of 1 electronic charge unit separated from an equal and opposite charge by a distance of 1×10^{-10} m, the dipole moment has the value 1.6×10^{-29} C-m.

The total molecular dipole moment of a molecule is in general the vector sum of various local electric dipoles that exist in the molecule. These local electric dipoles are of two types that we will call bond dipoles and lone pair dipoles. Local bond dipoles are generally larger in magnitude, hence more important, than lone pair dipoles.

They are thus the major contributors to the total molecular dipole moment, and we will discuss them first. A bond between two atoms, X and Y, in which the bond pair is unequally shared is called a polar bond. A polar bond gives rise to a local bond dipole. Unequal sharing occurs when X and Y have different electronegativities; the shared electron pair is on average closer to the more electronegative atom.

Consequently, the local bond dipole points along the bond, from the less electronegative atom (positive end) to the more electronegative atom (negative end). Assessment of the polarity of a bond is simple: if the two atoms involved in the bond are different either in identity or environment, the bond is polar.

Only bonds between identical atoms in identical environments are non-polar. A heteronuclear diatomic molecule such as HCl provides an example of a local bond dipole. Heteronuclear means that the two atoms of the molecule are different. Chlorine is more electronegative than hydrogen, so the bond electrons between them are closer to Cl than to H. This makes the Cl atom slightly negative and the hydrogen atom slightly positive. The H-Cl bond is polar, with the electrons biased toward Cl. This polarity gives rise to a local bond dipole that points along the bond from H (positive) to Cl (negative).

The second type of local electric dipole is the lone pair

dipole, which exists at any atom having one or more lone electron pairs. A local lone pair dipole points, as expected, from the nucleus of the atom (positive) to the lone pair (negative). Although strictly speaking there is a lone pair dipole generated by each lone pair in a molecule, the major contribution to the total molecular dipole is made by lone pairs on the central atom. Consequently, in examining a molecule for local lone pair dipoles, we will consider only lone pairs on the central atom, ignoring those on terminal atoms.

In addition to its local bond dipole, HCl contains three local lone pair dipoles, one associated with each of the lone pairs on the chlorine atom. The total molecular dipole moment of a molecule is the vector sum of all local electric dipoles (bond dipoles and central atom lone pair dipoles). The total molecular dipole moment of HCl is 0.36×10^{-29} C-m. This is due primarily to the bond dipole; the lone pair dipoles make a minor contribution, as is generally true.

Assessment of the total molecular dipole of HCl is quite simple, because there are only two atoms. For larger molecules, we will use the following systematic approach.

- First, examine each bond in the molecule for polarity. For each polar bond, draw a vector along the bond from the positive atom to the negative atom to represent the local bond dipole.
- Combine the local bond dipoles using vector addition methods to obtain the net bond dipole vector. The net bond dipole will usually be the major contributor to the total molecular dipole moment.
- Next, look for local lone pair dipoles on the central atom. Draw a vector from the atom symbol to each lone pair on the central atom. These are the local lone pair dipole vectors.
- Combine the central atom local lone pair dipoles using vector addition methods to obtain the net lone pair dipole vector.
- Combine the net bond dipole vectorially with the net lone pair dipole. The result is the total molecular dipole moment vector.

To illustrate the system, we consider the water molecule, which according to spectroscopy (and VSEPR) is bent, with an angle of 104°. Oxygen is more electronegative than hydrogen, so the O-H bonds are polar with the electrons closer to the oxygen than to the hydrogen atom. Each O-H bond in the water molecule therefore has a local bond dipole that points from H to O. The bond dipoles are along with their vector addition to produce the net bond dipole.

Because the water molecule is bent, the local bond dipoles add together to produce a non-zero net bond dipole that points from the centre of the (imaginary) line connecting the hydrogen atoms (positive) to the oxygen atom (negative). Next, we find two lone pairs on the central oxygen atom. Each lone pair generates a lone pair dipole pointing from the oxygen nucleus to the lone pair.

The result is a non-zero net lone pair dipole that points from the oxygen atom (positive) to the centre of the (imaginary) line connecting the lone pairs (negative). The net lone pair dipole and the net bond dipole point in the same direction. is the vector addition of the net bond and lone pair dipoles to produce the total molecular dipole moment of water.

The magnitude of the total molecular dipole moment is 0.615×10^{-29} C-m. This substantial dipole is a major factor in the ability of water to dissolve a variety of substances, notably including ionic compounds.

Although bonds between different atoms are polar to at least some extent, as we saw above, many highly symmetrical molecules have no net dipole moments due to exact cancellation of the bond and lone pair dipoles. An example is the CO_2 molecule, which is linear.

Each C-O bond is polar. Each therefore has a local bond dipole pointing directly along the bond from carbon (less electronegative) to oxygen (more electronegative).

However, the two equal local bond dipoles point in opposite directions and cancel exactly. The central carbon atom has no lone pairs, so no lone pair dipoles need be considered. Thus the molecule as a whole has coincident centers of + and–charge. Dipole moments contribute to the attractive forces that molecules exert on one another. To understand the magnitudes of these forces, it is essential that we be able to determine whether or not a molecule possesses a total molecular dipole moment.

Example. Which of the following molecules have (total molecular) dipole moments? SO_3, O_3, CCl_4, NH_3, PF_5.

Solution. Molecular polarity is assessed from molecular shape, which is in turn assessed from the Lewis structure. The method developed earlier leads to the following Lewis structures and shapes for these species. Lone pairs on the terminal atoms are not shown.

SO_3. Each S-O bond is polar, with a local vector bond dipole pointing from S (less electronegative) to O (more electronegative. All S-O bonds are the same when resonance is taken into account.

The three local bond dipoles thus point to the corners of an equilateral triangle. When added vectorially, the result is zero. There are no lone pairs on the central S atom, so local lone pair dipoles need not be considered. There is no net molecular dipole; SO_3 is non-polar.

O_3. Although the bonds in this molecule connect pairs of oxygen atoms, the two oxygen atoms of a bond are not in identical environments. Considering the bond between oxygens 1 and 2, we see that oxygen 1 is bonded only to oxygen 2, whereas oxygen 2 is bonded to oxygen 1, but in addition to oxygen 3. Oxygens 1 and 2 are in different environments, so the bond between them is polar. The difference between oxygens 1 and 2 is evident from their formal charges, which are 0 and 1+ in the resonance form shown.

The formal charges also enable us to assign the direction of the local bond dipole, which points from oxygen 2 to oxygen 1.

The molecule is bent, so the two identical local bond dipoles do not add to zero. They cancel horizontally, but not vertically. The local lone pair dipole pointing from oxygen atom 2 to its lone pair only partially cancels the vertical net bond dipole (recall our statement that in general bond dipoles are more important than lone pair dipoles). Ozone has a net molecular dipole running from the centre oxygen through the centre of the molecular angle.

CCl_4. From the Lewis structure for CCl_4, a perfect tetrahedral shape is expected (like methane, CH_4). The local bond dipoles are non-zero, with the negative end toward Cl, but cancel when added vectorially. The central carbon atom has no lone pairs.

There is no molecular dipole moment. The cancellation of local bond dipoles may not be obvious here because of the three-dimensionality of the figure. However, the following argument may be used. Molecules having total molecular dipole moments are inherently less symmetric than molecules with zero molecular dipole moments.

For example, if you were to orient the ammonia molecule with the nitrogen lone pair pointing down rather than up, it is easy to tell that the molecule has been moved. In contrast, if you were to orient carbon tetrachloride with any of the three other chlorine atoms at the top, the molecule would look exactly the same as it does above; you would not be able to tell that it had been moved! Thus it has no dipole moment.

NH_3. NH_3 has a trigonal pyramidal shape. The bonds are polar, with the negative end of the local bond dipole toward nitrogen. These will combine to give a non-zero net bond dipole colinear with an axis passing through the nitrogen atom and the centre of the equilateral triangle defined by the H atoms.

This is reinforced by the nitrogen local lone pair dipole. The molecule therefore has a molecular dipole moment with the nitrogen end of the dipole negative.

PF_5. PF_5 is trigonal bipyramidal. The bonds are very polar because F is the most electronegative element. However, the two axial local bond dipoles oppose and cancel, and cancellation occurs in the equatorial plane as well, exactly as in SO_3 above.

There are no lone pairs on the central atom. There is no molecular dipole moment.

REASONING BY ANALOGY: THE ISOVALENT FRAGMENT APPROACH TO STRUCTURE.

An approach to molecular structure based on analogical thinking. An oxygen atom has 6 valence electrons, so it requires 2 to achieve the octet. Suppose it gains 1 electron by forming a bond to hydrogen, giving the species H-O×. Here×represents the remaining odd electron on the oxygen atom that must pair with a second electron in bond formation.

Because the species H-O× requires only 1 electron more, it is analogous to a halogen atom, for example, Cl×. Again,×represents the seventh chlorine electron, the one that is shown unpaired in the dot structure for the atom. Since the O atom in H-O× is capable of forming one bond, like Cl×, OH× and Cl× are isovalent (iso = same, so same valence). Further, we will call them fragments because they do not yet have electron octets.

Finally, they are 1-bond fragments because they need form only one additional chemical bond. Combining terms, we will refer to HO× and Cl× as 1-bond isovalent fragments. To emphasize the 1-bond aspect of these fragments, we will write them as OH- and Cl-, where the dash represents an incipient bond, not a negative charge. You should convince yourself that the following species are also 1-bond fragments, isovalent with HO- and Cl-:

H_2N–
H_3C–
Br–
H_2P–
H_3Si–
H–

Note: that the number of atoms attached to the central atom in these fragments increases by 1 for each 1-unit decrease in the group number of the central atom. Compatible fragments can be united to give the central atom of each fragment the octet. Thus two 1-bond fragments can be united. We therefore expect the existence of molecules in which two of the fragments from those listed above are combined. In fact, all possible combinations of the 1-bond fragments above actually exist and are (within the framework of the fragment idea) related, or analogous. The following structures are illustrative:

Cl – Cl	HO – OH	HO – Cl	H_2N – OH	H_3C –OH
Chlorine	Hydrogen Peroxide	Hypochlorous Acid	Hydroxyl Amine	Methyl Alcohol

In similar fashion, the oxygen atom itself can be viewed as a 2-bond fragment, represented O=, because by forming 2 bonds it can achieve the octet. The following species are 2-bond fragments isoelectronic with O=:

HN =
H_2C =
S =
$(CH_3)_2C$ =

2-bond fragments are compatible; thus two of them can be united to give the central atom of each fragment the octet. The following molecules can be viewed as the union of two 2-bond fragments:

O = O	Dioxygen
$H_2C = CH_2$	Ethylene
$CH_2 = O$	Formaldehyde
$(CH_3)_2C = O$	acetone

Further, a 2-bond fragment is compatible (can unite) with two 1- bond fragments, as in the following molecules:

OH_2
NH_3 (HN = plus 2H-)
CH_4 (H_2C = plus 2H-)

Finally, the nitrogen atom is a 3-bond fragment, requiring 3 electrons to achieve the octet. It may be represented as N(3). Isoelectronic 3-bond fragments are

HC(3)
(C(3))-
$(O(3))^+$
P(3)

Combination of such fragments gives the following molecules or ions:

HC(3)CH acetylene

C(3)O carbon monoxide
N(3)N dinitrogen
$N(3)O^+$ nitrosyl cation

A 3-bond fragment is compatible with three 1-bond fragments. You should generate some molecular formulas by combining 3-bond and 1-bond fragments.

A single 2-bond fragment may be viewed as equivalent to two 1-bond fragments (X= equivalent to 2 Y-), and may replace them in a formula. Thus phosgene, $COCl_2$, may be "generated" from carbon tetrachloride, CCl_4, by replacing two 1-bond Cl fragments with the 2-bond fragment, O=. CO_2 follows from phosgene by a similar replacement of the remaining Cl^- fragments.

Similarly, X(3) is equivalent to 3 Y-. It is quite easy to go from a known Lewis structure to several others using this analogical approach. The generation of a family of sulfur compounds. Successive replacement of two 1-bond F fragments with the 2-bond fragment, O=, leads to sulfur trioxide via two oxofluorides. Replacement of O= by another 2-bond fragment, HN=, leads to the Lewis structure for $S(NH)F_4$. Finally, replacement of three 1-bond F with one 3-bond fragment, N, generates the final molecule, SF_3N.

THE IMPORTANCE OF MOLECULAR STEREOCHEMISTRY

The importance of molecular shape cannot be overstated. It is the shape of the simple water molecule that determines its physical and chemical properties, and in particular makes it suitable for sustaining life. Shape has far reaching consequences in chemistry, biology, and the technology of materials.

It is probably not an exaggeration to say that virtually all important biological processes are grounded in molecular shape, and, in particular, the ability of molecules to recognize one another based on shape.

A familiar and extremely important example is found in the the double helical structure of the DNA molecule, responsible for the transcription of genetic information in all living organisms. The double helix is maintained because molecules recognize one another's shapes. The guanine and cytosine molecules engage in three attractive interactions that hold them in specific positions relative to one another. In biochemical jargon, this is called a base-pairing interaction; it occurs in the gap between the two DNA strands.

In 3-21b, adenine and thymine are shown in their base-pairing engagement, this time involving two shape-specific interactions. Clearly, molecular shape plays a crucial role in enabling these life-sustaining interactions.

Chemists are actively conducting research in the area of molecular recognition; that is, the manner in which molecules recognize each other based on shape. This area of research promises the development of new and useful polymers and plastics; and mechanical and electrical devices that can be used in the engineering of extremely small (nanoscale) machines and computers.

It will also help to clarify the details of many biochemical processes that are currently only vaguely understood. Among these are the interaction(s) between enzymes and their substrates that lead to catalysis of substrate conversion to products; the role of retinal in vision; the mechanism of smell; and the mechanism of nerve transmission.

We will have more to say about the nature of interactions between molecules. For now, it is important only to realise that such interactions are shape-driven.

Example. Application of structural concepts to an important problem: Why is the amide linkage in proteins planar?

Solution. Proteins are polymers constructed from amino acid monomers. This mean that single amino acids are strung together like snap beads to create an extremely long molecule called a protein, or polypeptide. There are some 20 amino acids used in making proteins, all having the basic structure shown in Figure. In the figure, R is a generic representation for a one-bond fragment that is attached to the central carbon atom. R is different for each distinct amino acid. Two amino acids link via reaction below, in which the nitrogen atom of the NH_2 group of one amino acid becomes bonded to the carbon atom that is doubly bonded to oxygen in the second amino acid.

$$NH_2\text{– CHR– COOH} + H_2N\text{– CHR'COOH} \rightarrow H_2N\text{– CHR–C(=O)–NH–CHR'COOH} + H_2O$$

Because water is produced as a result of the linkage, the process is called a condensation reaction. The group of atoms shown in Figure, consisting of the carbon and oxygen of the left amino acid and the nitrogen, hydrogen, and central carbon of the right amino acid, is collectively referred to as the peptide link. It is found experimentally that all of these atoms lie in the same plane—i.e., the peptide link is planar. It is our goal to provide a rationale for this observation in terms of the concepts of Lewis structure and VSEPR.

An acceptable Lewis structure for the peptide link is shown in Figure. This is the structure arrived at by applying the systematic procedure, and minimizing formal charges. The structure places three electron groups around the peptide carbon atom. All of these groups are used to bond atoms, so the stereochemistry at this carbon atom is trigonal planar. The structure places 4 electron groups around the nitrogen atom, one of them a lone pair.

This leads us to expect tetrahedral electron group distribution, and trigonal pyramidal stereochemistry at the nitrogen atom. However, this stereochemistry at the nitrogen atom would prevent some of the atoms in the peptide grouping from lying in the same plane with carbons 1 and 2 and the nitrogen atom. In particular, C3 and the proton attached to N could not simultaneously lie in the C1-C2-N plane in any arrangement of the molecule. How then do we explain planarity?

With some thought, we realise that there is another acceptable resonance form involving the N, C2, and O atoms of the peptide link. This is obtained by simultaneously moving the N lone pair in to form a double bond with C2, while moving the double bond pair between C2 and O out onto the oxygen atom as a lone pair.

This form, shown in Figure, is less acceptable than that

in d because it places formal charges of 1+ and 1- on N and O, respectively. However, these are not unusual formal charges for these atoms, and they are not in conflict with electronegativities. The most important aspect of the structure in 3-22e is that there are now 3 electron groups around the nitrogen atom! Judging from this resonance form, trigonal planar stereochemistry is expected at N, and the planarity of the atoms C2, N, H, and C3 is required. Further, the double bond between N and C2 also requires C1 and O to be in this same plane.

The planarity of the peptide link is understandable. The key question is this: when two resonance forms place different numbers of electron groups around an atom, what is the stereochemistry at that atom? Based on our earlier discussion of resonance, we know that the molecule is not rapidly switching back and forth between the two forms.

Rather, there is a single molecular structure that may be considered as an average of the resonance forms. And, in point of fact, the actual stereochemistry at N is trigonal planar. We will find the following rule to be useful in all situations like this: When the reasonable resonance forms for a molecule place different numbers of groups around a particular atom, the stereochemistry at that atom is consistent with the smallest number of electron groups in any resonance form.

Our final example teaches us an extremely important lesson. The principles of structure and stereochemistry that we developed from and applied to small molecules apply equally as well for large molecules. Thus the carbon, nitrogen, oxygen, and sulfur atoms in proteins follow the same bonding and stereochemical rules that apply in small molecules. In fact, the behaviour and function of biomolecules is a consequence of the adherence to these rules.

Chapter 4

Interaction of Light and Matter

THE QUANTIZED ATOM

A very interesting fact about atoms has been known since the middle of the 19th century. When atoms are irradiated with white light (light consisting of photons of all wavelengths), they absorb only certain definite frequencies of the light, not all frequencies. Atoms of different elements absorb different frequencies. How do we interpret this? We have seen that the frequency of light is a measure of its energy. Thus atoms absorb only light of certain energies.

This implies that the atom is only allowed to have certain energies. We now know that the electric field associated with the light photon interacts with the electrons of the atom. When a photon is absorbed by an atom, the energy of the photon is used to cause an electron of the atom to make a transition ("jump") to a higher energy state within the atom.

The photon that was absorbed disappears. Only certain energy states are allowed to the electrons. That is why only photons of certain frequencies (energies) can be absorbed. The photon energy is exactly the same as the energy *difference* between 2 allowed electron states. We write this as

$$DE = E(\text{state } 2) - E(\text{state } 1) = hn$$

This equation has been tremendously important in the development of our understanding of the detailed structures of atoms and molecules. Virtually everything we know about the atom, its internal structure, and the structures of molecules containing several atoms, is a result of studying the interaction of light with the atom or molecule. For you to have a qualitative understanding of this interaction is therefore very important.

THE ABSORPTION OF LIGHT BY ATOMS AND MOLECULES

We begin this section by stating some very important ideas, which will then be illustrated in subsequent discussion. The ideas are these:

- Light consists of oscillating electric and magnetic fields.
- Because nuclei and electrons are charged particles, their motions in atoms and molecules generate oscillating electric fields.
- An atom or molecule can absorb energy from light if the frequency of the light oscillation and the frequency of the electron or molecular "transition motion" match. Unless these frequencies match, light absorption cannot occur. The "transition motion" frequency is related to the frequencies of motion in the higher and lower energy states.
- By measuring the frequencies of light absorbed by an atom or molecule, we can determine the frequencies of the various transition motions that the atom or molecule can have. Thus we can use light absorption to probe the dynamics of atoms and molecules!

These ideas are the basis for the techniques of *spectroscopy*, the study of the interaction of light and matter. The word means "measurement of spectrum," where the spectrum of a substance is the array of frequencies of light absorbed by its atoms or molecules.

You should already be familiar with the idea of energy absorption through matching frequencies. This is exactly the way in which you transfer energy to a child on a swing. You push at a frequency that matches that of the swing. Energy is transferred successfully, making the swing go to a higher energy state (higher amplitude), only if the push is made at the right time. If the frequency of pushing does not match the frequency of swinging, very little if any energy is transferred. In fact, the swing may actually lose energy if you try to push it while it is moving back toward you. This idea operates in the same way when light and matter interact. It is complicated only by the fact that, in contrast to the swing, molecules can undergo more than one type of motion.

Molecular Motions and Energy

There are many ways in which a molecule can move, and each motion contributes to the energy of the molecule. *Molecular motion is quantized* — that is, each motion can occur with only certain energies. A molecule may gain or lose energy in one of these motions by absorbing or emitting a photon of light. The photon has energy that matches the allowed energy gain or loss associated with the motion.

By measuring the energies of the photons absorbed by the molecule, we obtain its spectrum. The spectrum provides information about the motions of the molecule. We begin with an expression for the energy of a molecule, written as a sum of the energies of the various motions possible to it. Each motion will then be described, along with the region of the electromagnetic spectrum with which it interacts. The total energy of a molecule may be written as follows:

$$E_{total} = E_{spin} + E_{translation} + E_{rotation} + E_{vibration} + E_{electrons} + E_{nucleus}$$

NUCLEAR MAGNETIC RESONANCE—ESPIN.

The nuclei of some (but not all) atoms behave as if they were rotating, or spinning, about an axis passed through them, much as a top spins about its central axis. This is quite easily visualized in terms of the particle nature of the nucleus; it is less easily visualized in terms of the wave picture, which by now we know is another face of nature that is manifested in the atomic/molecular realm. Spin of nuclei is similar to electron spin.

Examples of atoms whose nuclei spin are hydrogen, fluorine, the 13 isotope of carbon, and the 15 isotope of nitrogen. The spin of the hydrogen nucleus, because hydrogen atoms occur very frequently in molecules and are therefore very useful in determination of structure. When placed in a strong magnetic field, the spinning motion of the hydrogen atom gives rise to two energy levels.

The spin can be promoted from the lower to the higher level by interacting with and absorbing energy from electromagnetic radiation in the radio frequency (RF) region of the spectrum. A typical radio photon has a frequency of 10^8 s^{-1}, a wavelength of 3 meters (about ten feet), and an energy of 6.6×10^{-26} Joules. The separation between the two spin energy levels is a sensitive function of the molecular environment of the hydrogen atom in a molecule. Thus hydrogen atoms in different environments absorb photons of different energies.

By plotting the amount of energy absorbed by the spinning nuclei versus the frequency of the RF radiation applied to the molecule, we obtain the *nuclear magnetic resonance (NMR) spectrum* of the molecule. This spectrum provides information about the chemical environment of the spinning hydrogen nucleus, and can be used to deduce the atomic bonding patterns in the molecule.

NMR has now been developed to such a level of sophistication that it can be used to determine the 3-dimensional structures of huge protein molecules in solution. Recently, nuclear magnetic resonance spectroscopy has found medical applications, in which context it is known as magnetic resonance imaging (MRI). MRI utilizes the spinning motions of atomic nuclei to provide a map of the internal structure of human body tissue.

We will now discuss the nature of NMR spectra in a systematic way, beginning with spectra of simple molecules, and gradually escalating in complexity. As we proceed through the discussion, we will introduce some of the jargon used by practicing chemists when discussing NMR. As a first bit of jargon, we will henceforth refer to hydrogen atoms as *protons*. This is appropriate, because it is the nucleus of the hydrogen atom that is spinning, and this nucleus indeed consists of a single proton.

We begin by considering the NMR spectrum resulting from the molecule. The molecule has a single proton on the right (on carbon 1), and a pair of protons on the left (on carbon 3). Carbon atoms 1 and 3 are separated by a carbon atom attached to two chlorine atoms. The single proton on the right is attached to a carbon atom that is attached to 2 chlorine atoms.

In contrast, each proton on the left is attached to a carbon atom that is attached to one chlorine atom and a second hydrogen atom. The environments of the protons on carbon 1 and carbon 3 are clearly different. The energy level spacing of the two types of protons should thus be different, and two signals, one for each type of proton, should appear in the NMR spectrum.

Several aspects of this spectrum are important. First, the single sharp signal at exactly 0 ppm is not due to the subject molecule, but is instead due to a reference compound named *tetramethylsilane*, or TMS. By general agreement, the positions of NMR signals are reported relative to the position of the signal for TMS. Second, the unit used to measure resonance position, ppm (parts per million), is not a unit of energy. It has no units at all. It may be regarded as a measure of the shift in position between the TMS resonance and a particular resonance in the compound of interest.

It is usually called the *chemical shift*. The chemical shift indicates the chemical environment in which the spinning nucleus is found, as we will see shortly. Ignoring the TMS signal, then, which is present in all NMR spectra, we see that there are indeed two signals. Interestingly, the sizes, or intensities, of the signals are different. The right signal appears to be twice as tall as the one on the left.

This suggests that the right signal is due to the pair of protons on carbon 3, and the left signal to the single proton on carbon 1. It also appears that there is a relationship between the position of a signal in the spectrum and the number of chlorine atoms in the immediate environment of the proton(s) giving rise to the signal. We summarize our observations of the simple spectrum in Figure in terms of three rules:

- ***Rule:*** Each unique proton environment generates one signal.
- ***Rule:*** The intensity (size) of an NMR signal gives a measure of the number of equivalent atoms responsible for the signal. A two-proton signal is twice as large as a one-proton signal.
- ***Rule:*** The greater the number of electronegative atoms near a proton, the further to the left of TMS the proton signal will be.

Rule 3 requires some comment. In Figure, the signal due to the hydrogen atom on carbon 1 is shifted further away from TMS than is the signal of the hydrogens on carbon 3. In NMR jargon, the left signal is said to be further *downfield* (meaning further to the left) than the right signal. The reason for this is that the carbon-1 proton is close to two electronegative chlorine atoms. These pull electrons away from carbon 1, which in turn pulls them away from the hydrogen atom. The less electron density a hydrogen atom has around it, the further downfield its NMR resonance occurs. This type of consideration helps us to assign signals to particular hydrogen atoms in molecules.

Figure shows a molecule that appears simpler than the molecule in Figure, but is in fact somewhat more complex in NMR terms. The molecule in Figure can be obtained from that in Figure by removing the central carbon atom, bringing the proton-bearing carbons together.

Clearly the carbon 1 proton is still in a different environment than are the pair of carbon 2 protons, so we might expect the NMR spectrum of this new molecule to be very similar to that of the three-carbon molecule. Specifically, we might expect two signals, one twice as intense as the other, with the less intense one downfield. The spectrum is shown in Figure.

Although there are two signals, in roughly the expected locations, the signals appear to be split into subsignals. The upfield signal now consists of two, and the downfield signal of three, closely-spaced subsignals. Chemists call the upfield signal a *doublet*, and the downfield signal a *triplet*, to indicate the number of subsignals. Apparently, there is a complication showing up here that is not present in the spectrum in Figure. As we will see, this complication is what makes NMR so useful for determination of structure. The splitting of signals seen in Figure occurs because protons on neighboring atoms "feel" each other's spins.

A spinning nucleus is, in fact, a tiny magnet. When it is close enough to another proton, the second proton feels the small magnetic field of its neighbour as a tiny perturbation on the large magnetic field applied to the sample. The effect of this proton-proton interaction is to cause small changes in the energy level spacings of the proton that show up as the splittings that we see in Figure. Interestingly, the number of subsignals into which a signal is split appears to be one greater than the number of neighboring protons. We can now add a couple of new rules to our list:

- ***Rule:*** Protons on neighboring atoms interact with each other, and split each other's signals.
- ***Rule:*** The number of subsignals in a *multiplet* is one more than the number of neighbour protons.

Clearly the splitting of signals due to neighboring atoms (called *spin-spin splitting* in the jargon of the chemist) is of potentially great value in assigning signals and in determining structure; it tells us the number of protons on the atom next door to the atom bearing the protons being observed.

To test our rules, we consider the molecule in Figure. Again we see only two different proton environments here. The two carbon-1 protons are in one environment; the three carbon-2 protons are in a different environment. We expect two signals, one of total intensity two and the other of total intensity three. The signal of intensity two should be further downfield because the carbon 1 protons have an electronegative chlorine atom in their neighborhood. We expect the downfield signal to be split to a quartet due to interaction with the three protons on carbon 2; and the upfield signal to be split to a triplet via interaction with the two protons on carbon 1. The spectrum is in Figure, and appears much as we expected it to. The NMR rules seem to work fine.

We consider one final example, the NMR spectrum of isobutyraldehyde, which will reveal a couple of other useful subtleties of NMR spectra. The molecule of interest and its NMR spectrum are shown in Figure and Figure. The protons labelled "a" in the figure experience the same environment (this is clear from a model of the molecule), and are expected to give one signal.

This signal should be the furthest upfield because these protons are furthest removed from the electonegative oxygen atom. Because there is a single proton, "b", on the neighboring carbon atom, the signal for protons "a" should be a doublet with total relative intensity 6. This is clearly seen in the spectrum. Proton "c" should give the furthest downfield signal because it is attached to an oxygen-bearing carbon atom.

This, too, is expected to be a doublet due to interaction with proton "b". Although the splitting of the proton "c" signal is small, it is discernible in Figure. Note that the signal from "c" is very far downfield, much more so than the signals influenced by chlorine atoms in previous examples. This results from the very high electronegativity of the oxygen atom, which substantially depletes the electron density of the attached carbon atom.

Finally, we examine the signal for proton "b". This is our first encounter with a proton that interacts with two other sets of protons, and it is important to understand how the splitting works in a situation like this. Figure shows a magnified view of the signal for proton "b", which occurs at a chemical shift of 6.4 in. The signal shows a total of 14 lines, which is many more than we have encountered in a multiplet up to this point.

The explanation for all of these lines runs as follows. First, the signal for "b" is split to 7 lines by the interaction with the 6 "a"-type protons on neighboring carbon atoms. Then, each component of the 7-line pattern is split to two lines by the interaction with proton "c". The total is 14 lines. In terms of this interpretation, Figure clearly shows a 7-line pattern, each component of which is split into two lines. This pattern is called a doublet of septets, to indicate that it results from interaction of "b" with two distinctly different types of protons. On the basis of the isobutyraldehyde example, we are in a position to state two additional rules.

- ***Rule:*** A proton on an oxygen-bearing carbon is shifted very far downfield.
- ***Rule:*** A proton with "n" neighboring protons on one side and "m" on the other will give a signal split into (n+1)(m+1) components.

Example. The structure of propionaldehyde, C_3H_6O, is shown in Figure. Predict what the NMR spectrum of propionaldehyde should look like.

Solution. The spectrum should show 3 signals of relative intensities 3 to 2 to 1. The signal of intensity 3, due to the hydrogen atoms labelled "a" in the figure, will be furthest to the right because these atoms are far removed from the oxygen atom. The signal should be split into 3 components from the effect of the spins of the two H atoms on the middle carbon.

The signal of intensity 1, due to hydrogen "c", is furthest left because this hydrogen atom is near the oxygen atom. This signal also will be split into 3 components due to the effect of the spinning motions of the two H atoms on the middle carbon atom. Finally, the signal due to hydrogens "b" will appear at an intermediate position, with relative intensity 2. It will be split to two components by hydrogen "c"; and each of these will be split to 4 components by hydrogens "a". This signal will consist of 8 components altogether.

Chemists have invented a shorthand system for presenting the information in an NMR spectrum without having to have the spectrum itself (i.e., the picture) available. Each signal in the spectrum is described by stating, first, its multiplicity (the amount of splitting); second, the chemical shift in parts per million downfield from TMS; and third, the relative intensity of the signal, in terms of number of identical protons giving rise to the signal. The propionaldehyde spectrum is described as follows in this shorthand system:

triplet, d = 1.1 (3H)
quartet of doublets, d = 2.45 (2H)
triplet, d = 9.7 (1H)

The Greek letter, d, symbolizes the chemical shift in units of parts per million. This brief description is sufficient to allow us to visualize and, if desired, even draw the spectrum, and to interpret it in terms of structure.

In this section, we have considered a number of NMR spectra and from them have developed a simple set of rules for their interpretation and prediction. It is important to realise that, although these rules are very useful in understanding the NMR spectra of relatively simple molecules, the spectra of many molecules involve complexities that our simple rules are unable to handle.

Spin-spin Splitting—A Closer Look

To this point, we have discussed the most basic aspects of nuclear magnetic resonance spectroscopy. We have acknowledged the existence of spin-spin splitting, but have not said much about its molecular origins, and have not explained the origin of the number of subsignals in a multiplet. In this section we will pursue these matters in a bit more depth. As a basis for discussion, consider a molecule in which there are three protons on carbon 1 and two protons on the neighboring carbon atom 2.

Our rules tell us that the signal due to carbon-1 protons will be split to a triplet, and that the signal due to carbon-2 protons will be split to a quartet. Why a triplet? Why a quartet? Why do the subsignals of the triplet have relative intensities 1:2:1, while those of the quartet have relative intensities 1:3:3:1? What is the significance of the *magnitude* of the splitting between the components of a multiplet?

Focus first on the splitting of the signal due to the carbon-1 protons. The spinning protons on carbon 2 behave like tiny magnets. The magnetic field is generated by the spinning negative charge, which is much like an electric current. The direction of the magnetic field generated by a carbon-2 proton depends on its direction of spin (clockwise or counterclockwise). Either direction is equally probable for each of the three carbon-2 protons. Given this, it is easy to appreciate that there are 3 possible combinations of spin directions for the 2 protons on carbon 2.

Both can be spinning clockwise, in which case their magnetic fields add together and reinforce the large laboratory magnetic field. One can be spinning clockwise and the other counterclockwise, in which case their magnetic fields cancel. This arrangement can occur in two different ways because either proton 1 or proton 2 can have clockwise spin. The arrangement of one spin "up" (cw) and the other "down" (ccw) is thus twice as probable as the arrangement with both spins "up".

Finally, both protons can have spin "down", in which case their magnetic fields add together but oppose the applied laboratory field. The spin arrangements of the carbon-2 protons are perhaps better appreciated when presented pictorially. We conclude that the carbon-1 protons will feel three slightly different magnetic fields, one slightly larger than, one equal to, and one slightly smaller than the applied field. The middle possibility is twice as probable as the other two.

Now we are in a position to appreciate why the signal due to the carbon-1 protons is split to 3 subsignals of relative intensities 1:2:1. This is a simple consequence of the fact that the carbon-2 protons create 3 slightly different magnetic fields that can be experienced by the carbon-1 protons. One signal arises from each of these magnetic fields, with intensity proportional to the probability of the spin arrangement giving rise to the magnetic field perturbation caused by protons on carbon 2.

In similar fashion, we can understand why the signal due to carbon-2 protons is split to 4 subsignals. The 3 protons on carbon 1 can have their spins aligned in 4 different ways (all "up" (one possibility), two "up" one "down" (3 possibilities), one "up" two "down" (3 possibilities), and all "down" (again only one possibility). The carbon-2 protons will experience 4 slightly different magnetic fields with probabilities 1:3:3:1, so will be split to a quartet whose components have intensities reflecting these probabilities. The number of components in a multiplet and their relative intensities are thus seen to have a physically plausible explanation.

In general, as we have seen, "n" neighboring protons cause a splitting into "n+1" components. For reasons that we need not pursue, the relative intensities of the components resulting from splitting by any number of equivalent neighboring protons can be readily computed from Pascal's Triangle, in which a particular number is calculated as the sum of the numbers to the left and right of it in the row above.

We conclude our examination of spin-spin splitting by considering the magnitude of the separation between two adjacent signals within a multiplet. We can base the initial part of the discussion on the same molecule used above, in which there are three protons (call them protons A) on carbon 1 and two protons (protons B) on the neighboring carbon atom 2.

The magnetic interaction between protons A and B is called spin-spin coupling. Protons A are said to couple with (meaning that they interact with) protons B. The strength of the interaction determines the magnitude of the splitting between adjacent components of a multiplet.

The stronger the interaction, the larger the splitting. Further, spin-spin coupling between the two sets of protons A and B obeys a version of Newton's third law, that for every action there is an equal and opposite reaction. If protons A interact with protons B with a certain strength, splitting the B signal by a certain amount, then protons B interact with protons A with exactly the same strength, splitting the A signal by exactly the same amount.

The size of the splitting within a multiplet is called the *coupling constant, J*, and is usually measured in frequency units. We will not be concerned with measuring coupling constants, so do not worry at this point about the units. The importance of coupling constants for us will be this: the signals for two sets of interacting (coupled) protons show exactly the same coupling constant.

Thus it is often possible by looking at the NMR

spectrum to decide which signals in the spectrum are coupled. This is an additional aid in building the structure of a molecule from an analysis of its NMR spectrum.

To show how this works, we complicate our test molecule somewhat by adding a third carbon to which is attached a single proton, D. We expect the following coupling interactions within this structure:

- A with B only; the signal for A should be a 1:2:1 triplet
- D with B only; the signal for D should be a 1:2:1 triplet
- B with A and also with D; the signal for B should be a 1:3:3:1 quartet as a result of coupling with A. Each component of the quartet should be split to a doublet as a result of coupling with D.

  Suppose that the interaction between A and B is stronger than that between B and D. For discussion sake, we will assume that B and D couple with only 1/3 the strength of A and B. Thus if the coupling constant for interaction between A and B is 12 s^{-1}, then that for interaction between B and D is 4 s^{-1}. Then we can make the following additional statements about the appearance of the NMR spectrum:
- The splitting within the multiplet due to protons A (12 s^{-1}) will be 3 times bigger than the splitting within the multiplet due to protons D.
- The B multiplet will consist of 4 pairs of signals. The members of a pair will be separated by 4 s^{-1}; the corresponding members of two adjacent pairs will be separated by 12 s^{-1}.

The spectrum of our test molecule might look as shown, where we have assumed chemical shifts of 1 ppm, 2 ppm, and 4 ppm for the A, B, and D protons, respectively.

INFRARED SPECTROSCOPY—EVIBRATION

Infrared spectroscopy allows us to examine the vibrational motions of molecules. We will discuss these motions as one of several types of motion that the molecule as a whole can undergo.

Etranslation, Erotation, Evibration

The nuclear spinning motion discussed above involves only one atom of a molecule. The three motions to be discussed now — translation, rotation, and vibration — are motions of the molecule as a whole in space, involving concerted movements of all of the atoms. For illustration, we use a very simple molecule consisting of only 2 atoms. Such molecules are called *diatomic molecules.* Examples are O_2, H_2, HCl, and CO. A generic diatomic molecule is pictured in Figure.

First, the entire molecule may move through space in an arbitrary direction and with a particular velocity. This is called *translational motion* and with it we associate the translational kinetic energy of the molecule:

$E_{trans} = 1/2\ mv^2$ (m = mass of molecule; v = velocity of centre of mass of the molecule)

The translational velocity of the molecule has components along each axis of a Cartesian coordinate system, so

$E_{trans} = 1/2\ mv^2 = 1/2\ mv_x^2 + 1/2\ mv_y^2 + 1/2\ mv_z^2$, where v_x is the x-component of velocity, etc., and m is the mass of the molecule

The total translational KE of the molecule is made up of three parts, each representing the kinetic energy of the molecule along one reference direction. The molecule has 3 translational degrees of freedom (i.e., independent directions in which it can translate), one corresponding to each Cartesian axis.

A molecule travelling through space at a temperature of 300 K has translational kinetic energy of 6.2×10^{-21} Joules. However, the spacing between the allowed translational energies for an HCl molecule in a container 20 cm in length is on the order of only 10^{-40} J. This is such a small energy that transitions between translational kinetic energy levels are difficult to observe. At the present time, no spectroscopy is done involving these energy levels.

Second, the molecule may rotate about an internal axis. For example, rotation of a diatomic molecule is diagrammed in Figure. Like translation, a molecular rotation may be resolved into three mutually perpendicular components. The rotational kinetic energy of the molecule is

$$KE_{rot} = 1/2\ I_x w_x^2 + 1/2\ I_y w_y^2 + 1/2\ I_z w_z^2,$$

where I_x, I_y, and I_z are the moments of inertia (the moment of inertia is the rotational equivalent of mass) about the x, y, and z axes, and w_x, w_y, w_z are the angular velocities (in units of radians/s) about these axes. There are 3 *rotational degrees of freedom*, one for each Cartesian axis. An exception arises in the case of a linear molecule, for which one of the three axes is coincident with the molecular axis. The molecule has no rotational energy about this axis, because the corresponding moment of inertia is zero (i.e., no mass lies off the axis). Linear molecules have only two rotational degrees of freedom.

Rotation is a type of oscillation, in which the number of rotations per second is the frequency. If the molecule has associated with it a permanent electric field, then rotation of the molecule about some axis causes an oscillation in the electric field. If this oscillation matches in frequency the oscillation in the electric field of an impinging photon, the molecule can absorb the photon and go to a higher energy rotational state. Rotations of molecules at room temperature typically occur with energy on the order of 10^{-23} J.

Photons with frequencies near 10^{11} s^{-1} (wavelength, 0.03 m), in the microwave region of the electromagnetic spectrum, are required to interact with such motions. The rotational (or microwave) spectrum of a molecule gives information about the moments of inertia, from which can be extracted the bond lengths and bond angles in the molecule. Interpretation of rotational spectra is complex, and we will not be concerned with it at this time.

Finally, the molecule may vibrate. Chemically bonded atoms are like masses connected by a spring (the bond). The bonds can be stretched and compressed somewhat, and the bond angles can vary within certain limits. Motions of the molecule in which bonds are compressed and/or stretched, or in which bond angles get larger then smaller, but in which the centre of mass of the molecule does not move through space, are called *vibrational motions*. A diatomic molecule vibrates by repeated stretching and contraction of the bond joining the two atoms. The molecule has one *vibrational degree of freedom* in addition to 3 translational and 2 rotational degrees of freedom.

For a polyatomic molecule the number of vibrational degrees of freedom (called *vibrational modes*) is the total number of degrees of freedom minus those of translation and rotation. The total is 3N, where N = the number of atoms in the molecule.

This is true because each atom may independently move in any of three directions and therefore has 3 degrees of freedom available to it; the molecule has a total of 3N. The number of vibrational degrees of freedom is 3N-6 for a non-linear polyatomic molecule, and 3N-5 for a linear polyatomic molecule. The water molecule, with N = 3, has $3 \times 3 - 6 = 3$ vibrational degrees of freedom.

Vibrational motions occur with characteristic frequencies (just as does the motion of the playground swing). In addition, the vibration causes a change in the distribution of the electrons in the molecule, which may give rise to an oscillating electric field. If the frequency of impinging light matches the frequency of the vibrational motion, the molecule may absorb light photons (energy) and move to a vibrational state in which the amplitude of vibration is increased.

By measuring the frequencies of light absorbed, we may observe how rapidly the molecular vibrations occur! Since the frequencies of vibrations are directly related to the strengths of the chemical bonds (i.e., to the stiffness of the spring), vibrational frequencies give us information about chemical bond strengths.

Typically, molecules vibrate at between 10^{13} and 10^{14} times per second. Photons with frequency 10^{14} s^{-1} (wavelength 3×10^{-6} m) have energy 6.6×10^{-20} J. Vibrational motions therefore have frequencies that are about 10^3 times larger than the frequencies of molecular rotation and a million times larger than nuclear spin frequencies. Higher energy photons in the infrared region of the spectrum are required to excite the vibrations.

Infrared (IR) spectroscopy deals with transitions between vibrational energy levels in molecules; it is therefore also called *vibrational spectroscopy*. Because IR spectra are easily acquired and provide much structural information, we will discuss them in some detail. An IR spectrum is a plot of the energy of the infrared radiation (expressed either in micrometers (1 mm = 10^{-6} m) or wavenumbers, the reciprocal of wavelength in centimeters) versus the per cent of light transmitted by the compound.

Within the energy range between 200 and 4000 cm^{-1}, the vibrational spectrum of the molecule appears as a series of absorption bands of variable intensity. The absorption bands are the downward-facing spikes, or peaks. Note that when absorption occurs, the per cent of light transmitted by the sample decreases from near 100 to various smaller values. Each absorption band in the spectrum corresponds to a vibrational transition within the molecule, and gives a measure of the frequency at which the vibration occurs. For the water molecule, with three vibrational degrees of freedom, there are three sets of energy levels within which transitions may occur.

The spacing between energy levels depends upon the nature of the vibration. Each spacing requires a photon of different energy to cause the transition, so infrared photons of three different energies are absorbed by H_2O. The frequencies of the photons (and of the vibrations) are 3600, 3500, and 1650 cm^{-1}, respectively, for the asymmetric stretch, the symmetric stretch, and the bend.

All molecules vibrate, but not all vibrations interact with electromagnetic radiation. For a vibrational motion to absorb infrared electromagnetic radiation, it must produce a change in the dipole moment of the molecule. HCl, for example, with a centre of positive charge located near the H atom and a centre of negative charge located near the Cl atom, has a dipole moment. Moreover, the magnitude of the dipole moment changes as the bond stretches, so this vibration is able to absorb IR radiation and is said to be *IR active*. HCl thus exhibits an IR spectrum. The N_2 molecule, on the other hand, has no dipole moment. Furthermore, stretching the NN triple bond does not change the dipole moment.

Consequently, the molecular vibration is infrared inactive (cannot directly absorb IR radiation). There are many molecules that, although possessing no permanent dipole moment, still undergo vibrations that change the value of the dipole moment from 0 to some non-zero value. Consider the CO_2 molecule. This molecule has no permanent dipole moment, since the individual bond dipoles exactly cancel. However, when it undergoes a bending vibration its dipole moment changes from zero to some non-zero value.

This vibration produces a change in dipole moment and is therefore IR active. All three vibrations of the water molecule change the dipole moment and are IR active. Three bands appear in the IR spectrum. The requirement that a vibration must cause a change in the dipole moment in order for a molecule to absorb radiation is another situation in which an oscillating electric field in the molecule must match that in the radiation. Oscillation of the dipole moment generates such an oscillating electric field.

In addition to the *number* of distinct vibrations expected for a molecule, we can anticipate their *locations* in the spectrum. For a diatomic molecule, A-B, the wavenumber, n, of the infrared radiation absorbed by vibration of the molecule is given by

$$n_{AB} = n/c = 1/l = (1/2p)\ (k_{AB}/m_{AB})^{1/2}w$$

$$\text{here } m_{AB} = M_A M_B/(M_A + M_B)$$

k_{AB} is the force constant for the A-B bond and measures the bond strength; m_{AB} is called the reduced mass; and M_A and M_B are the masses of the atoms. This equation indicates that the heavier the atoms involved in the bond, the lower the absorption frequency, given a constant bond strength. Care is required in applying this equation, however. For the series HF, HCl, HBr, and HI, the IR absorptions occur at, respectively, 3958, 2885, 2559, and 2230 cm^{-1}. It is tempting to conclude that this is due to the increase in mass of the halogen atom. Calculation of m for each molecule reveals that the reduced mass actually changes very little.

The change in frequency is instead due to the decrease in bond strengh along the series. Equation suggests that, because the frequency of a vibration depends on the identities of the two bonded atoms and the strength of the bond between them, a particular type of bond should always absorb at more or less the same location in the infrared spectrum. This turns out to be true. Thus, for example, all molecules containing C-H bonds show absorptions in the region of 3000 cm^{-1}, corresponding to the stretching motions of these bonds.

To develop a feel for the manner in which structural changes in the framework of a molecule are reflected in the infrared spectrum, we will consider the IR spectra of a series of relatively simple molecules based on a chain of 5 carbon atoms. The first member of the series has formula C5H12, and is called pentane. The infrared spectrum of pentane is in Figure.

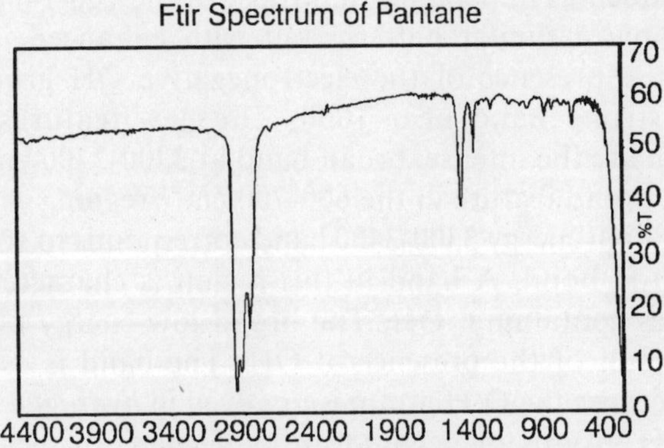

Fig. IR Spectrum of Pentane

The spectrum is relatively uncomplicated. There is a group of absorption bands in the region 2800-3000 cm^{-1} that correspond to C-H stretching vibrations; a pair of bands in the 1350-1500 cm^{-1} region, due to C-H bending vibrations; and a series of very weak bands in the region between 700 and 1400 cm^{-1}, due to various complex motions of the molecule that involve C-C stretches, C-C bends, and more complicated C-H motions. The most intense absorption in the spectrum is clearly that due to C-H stretching, because the largest dipole moment change arises from these motions.

The IR spectrum of 1-chloropentane, $C_5H_{11}Cl$. The number preceding the name of this compound indicates that the chlorine atom is attached to the first carbon atom in the chain. The spectrum is remarkably like that of pentane, with two obvious changes. First, the bands in the 700-1400 region, although they show the same pattern as in this region of the pentane spectrum, are more intense than those for pentane.

This is a result of the chlorine atom, which makes the molecule polar, resulting in a larger change in dipole moment during many of the vibrations. Thus intensity of absorption is increased. Second, there is a strong band at 650 cm^{-1} that is absent in the pentane spectrum. This band must be associated with the C-Cl stretching motion.

Next, we consider the spectrum of 1-pentanol, $C_5H_{11}OH$. Again, the bands in the 2800-3000 and 1350-1500 regions appear much as in pentane; and those in the 700-1400 region again show a similar pattern, but with enhanced intensity due to the presence of the electronegative OH group, and with a strong band near 1050. The new features of the spectrum are the intense, broad band at 3300-3400 cm^{-1}, and the very broad feature in the 600-700 cm^{-1} region.

The high energy 3300-3400 band corresponds to stretching of the O-H bond. A band in this region is characteristic of molecules containing -OH. The broad low energy feature is also a result of the presence of OH. The band is extremely broad because the O-H group is involved in *hydrogen bonding* with O-H groups of neighboring molecules. In most cases, molecules containing O-H groups are involved in hydrogen bonding to some extent, and produce a broad O-H stretching band in the IR spectrum. Hydrogen bonding, one of several types of *intermolecular force*.

Finally, we examine the spectrum for the pentane derivative, CH_3-C(=O)-CH_2-CH_2-CH_3, which is called 2-pentanone (the -one ending tells us that the molecule contains the C=O unit). The striking feature of this spectrum is the intense band at 1700 cm^{-1}, corresponding to the stretching vibration of the C=O group. All molecules containing the C=O group have a band in the region 1680-1750 cm^{-1} in the infrared spectrum. The presence of such a band is an indication of this structural feature.

This brief discussion of the IR spectra of pentane derivatives should give you an appreciation of the relationship between structure and the positions of absorption bands in the IR spectrum. Let's look at two further examples.

Example. The structure and infrared spectrum of acetone, C_3H_6O. What structural information can be determined from this spectrum?

Solution. The bands just above 2800 cm^{-1} are due to C-H stretching motions, which occur at high frequency because the H atom has such small mass. Any compound containing C-H bonds will have an absorption band in this region. The intense absorption band at 1730 cm^{-1} is due to the stretch of the C=O bond. C=O bonds invariably absorb in the region 1750-1680 cm^{-1}, independent of the other groups bonded to carbon.

The appearance of a band in this region in the spectrum of an unknown compound is diagnostic of a carbon-oxygen double bond. The bands at energy lower than 1500 cm^{-1} arise from bending motions of C-H bonds and stretching and bending motions of C-C bonds. As a general rule, it is difficult to assign bands in this region to particular bonds in the molecule.

Example. The structure and infrared spectrum of t-butanol, $(CH_3)_3COH$. What structural information is available from the spectrum?

Solution. The most prominent feature of the spectrum is the strong, broad band at about 3400 cm^{-1}. This is due to the stretching vibration of the O-H bond. The band is strong (i.e. intense) because there is a substantial change in dipole moment when the O-H bond stretches. It is broad as a result of hydrogen bonding, which is the interaction between the hydrogen atom of the O-H group of one molecule with the oxygen atom of the O-H group of another. Hydrogen bonding, an example of an intermolecular force.

Many molecules containing an OH group exhibit a similar vibrational band in the region 3200-3600 cm^{-1}. Next, the intense series of bands between 2800 and 3000 cm^{-1} is due to stretching motions of the various C-H bonds in the molecule. That they all occur at 3000 or less indicates that the hydrogen atoms are attached to carbon atoms that form single bonds only (rather than, say, double bonds).

Finally, the intense, somewhat broad band at 1200 cm^{-1} is due to the stretching vibration of the C-O single bond. The remaining bands cannot be assigned to particular bond types; instead, they are due to complex, interdependent vibrational motions of the molecule. However, the pattern of bands in the region below 1400 cm^{-1}, usually called the *fingerprint region*, is unique to a particular molecule, and can be used to identify it.

Functional Group Analysis and Fingerprinting

An *organic functional group* is a small group of atoms, bonded together in a characteristic way, that occurs in a large variety of compounds. Examples are C=O (carbonyl), C(=O)OH (carboxyl), and NH_2 (amine).

The functional group is often emphasized in writing the formula of a compound containing it. Thus R-COOH represents any molecule containing the carboxyl functional group. R is a generic symbol for an organic "radical", consisting of primarily carbon and hydrogen atoms bonded together, and in which one carbon atom is capable of forming one additional bond.

Organic functional groups differ from one another both in the strength of the bond(s) and in the masses of the atoms involved. For instance, the O-H and C=O functional groups contain atoms of different masses connected by bonds of different strengths.

According to equation, we therefore expect the O-H and C=O groups to absorb IR radiation at different positions in the spectrum. The presence of a strong, broad band between

3200 and 3400 cm^{-1} indicates the presence of an O-H group in the molecule, while the presence of a strong band around 1700 cm^{-1} confirms the presence of a C=O group.

For organic molecules, the infrared spectrum can be divided into three regions. Absorptions between 4000 and 1300 cm^{-1} are primarily due to specific functional groups and bond types. Those between 1300 and 900 cm^{-1}, the *fingerprint region*, arise from more complex vibrations in the molecules; and those between 900 and 650 cm^{-1} indicate benzene rings in the molecule.

Table. Important Functional Groups

Functional Group	*Wavenumber Range*
O-H	3400-3200
N-H	3300-3000
C(3)C-H	3300
=C-H	3100-3000
C(=O)-H	2700
C=O	1700
C=C	1680-1600
Benzene	1600
C-C-O	1260-1000
strong bands characteristic of benzene	900-650

We will briefly discuss the characteristic infrared absorption frequencies of a variety of common organic functional groups. While studying these, realise that the infrared spectrum of even a simple compound can be very complex. Some absorption bands are weak while others are strong. Some are close enough together to coincide or overlap. Interactions such as hydrogen bonding cause absorption bands to shift and broaden.

Accordingly, do not expect to be able to identify every absorption band in a spectrum. Rather, look for the presence or absence of absorption bands in their characteristic spectral regions to confirm the presence or absence of particular functional groups.

Alkanes. Alkanes are hydrocarbons containing only single bonds. The most prominent bands in infrared spectra of alkanes and from alkane portions of more complicated organic compounds are due to C-H stretching and bending. The symmetric and asymmetric C-H stretching frequencies are in the regions 3000 to 2900 and 2900 to 2800 cm^{-1} respectively, and are usually weak. Because most organic compounds contain several $-CH_3$, CH_2, and/or -CH groupings, there is often overlap of absorption bands in this region.

Thus the presence or absence of bands is taken to indicate the presence or absence of C-H bonds in the molecule.

The major bending modes of CH_2 and CH_3 groups appear at 1470-1420 and 1380-1340 cm^{-1}. The exact positions depend upon the nature of adjacent atoms. Interpretation is usually complicated by the presence of several bands of this type, or additional bands from other sources. The (usually) large number of C-C bonds in an organic molecule makes the C-C stretching vibrations in the 1300-1100 cm^{-1} region uninterpretable in most cases.

Alkenes. Alkenes are hydrocarbons containing at least one C=C double bond. When H is attached to a double-bonded C, C-H stretching bands generally appear in the region 3100-3000 cm^{-1}. Alkanes and alkenes can thus be differentiated. The C=C stretching frequency occurs in the 1675-1600 cm^{-1} region and varies with the atoms attached to the double-bonded carbons.

Table. Alkene Bending Absorptions

Alkene	*Wavenumber range, cm^{-1}*
$R-CH=CH_2$	1000-960 940-900
$R_2C=CH_2$	915-870
trans-RCH=CHR	990-940
cis-RCH=CHR	790-650
$R_2C=CHR$	850-790

Alkynes (or *acetylenes*) Alkynes are hydrocarbons containing a carbon-carbon triple bond, C(3)C). The C-H stretching vibration of *terminal acetylenes* (meaning that the last carbon in a chain is triple bonded to its neighboring carbon) generally appears at 3300 cm^{-1} as a strong sharp band. The CC stretching band is found at 2150-2100 cm^{-1} if the alkyne is monosubstituted (i.e., R-C(3)C-H) and at 2270-2150 cm^{-1} if disubstituted (i.e., R-C(3)C-R'). These absorptions are usually weak.

Aromatic (benzene) Rings. Aromatic C-H stretching absorption appears in the region 3100-3000 cm^{-1} and is readily distinguished from alkane, alkene, or alkyne C-H, allowing a reliable determination of the types of carbon-bound hydrogen in the molecule. The positions of aromatic C-H bending bands at 900-690 cm^{-1} depend upon the substitution pattern of the benzene ring as indicated in Table.

In the absence of other interfering absorptions, these strong, usually sharp bands can be used to distinguish postitional isomers of substituted benzenes. Sharp bands at ~1600 and ~1500 cm^{-1} are characteristic of all benzene compounds; a band at 1580 cm^{-1} appears when the ring is *conjugated* with a substituent. (The word conjugated means the substituent has a double bond one bond removed from the benzene ring).

Table. Absorptions of Benzene Rings

Substitution	*Wavenumber range (cm^{-1})*
mono	755-730 710-690
1,2-di	765-730 710-690
1,3-di	800-750 710-690
1,4-di	840-800
1,2,3-tri	800-760 740-700
1,2,4-tri	880-860 820-800

Alcohols. Alcohols are compounds containing the O-H functional group. The characteristic infrared band due to the O-H stretching vibration appears at 3650-3600 cm^{-1} in dilute solution. With increasing concentration in solution, or in pure liquids or solids, intermolecular hydrogen bonding broadens the band and shifts its position to lower wavenumber (3500-3200 cm^{-1}). The infrared spectrum of pure t-butyl alcohol shows the broad O-H stretching band at 3440 cm^{-1}. In addition to the O-H stretching vibration, alcohols exhibit O-H bending and C-O stretching transitions at 1500-1300 cm^{-1} and 1220-1000 cm^{-1}, respectively. Although C-O stretching bands occur in a spectral region where there are usually many other bands, they are easy to identify because they are intense (strong). They are often *coupled to* (i.e., move in concert with) C-C absorption bands and therefore exhibit splitting.

The exact location of the C-O bands depends on the degree of branching of the carbon atom that is attached to oxygen. Thus, as Table reveals, the position of the C-O absorption band can allow distinction between *primary*, *secondary*, and *tertiary* alcohols. In a primary alcohol, the OH group is attached to a carbon that is in turn attached to only one other carbon atom. In secondary and tertiary alcohols, the OH group is located on a carbon attached to 2 and 3 other carbons, respectively.

Table: C-O Stretching Vibration of Alcohols

Primary alcohols	1050 cm^{-1}
Secondary alcohols	1125 cm^{-1}
Tertiary alcohols	1200 cm^{-1}

Ethers. Ethers are compounds containing the C-O-C functional group. The asymmetric C-O stretching absorption of ethers appears in the region 1280-1050 cm^{-1}. As in alcohols, the position of this strong band is dependent on the nature of the attached groups.

Aldehydes and Ketones. The carbonyl, (C=O) stretching absorption is particularly important because it is present in a variety of functional groups including aldehydes, ketones, and carboxylic acids and their derivatives. Table shows the basic structures of compounds containing C=O.

Table: Carbonyl Containing Compounds

Name	*Structure*
Aldehyde	R-C(= O) – H
Ketone	R-C(=O) – R
Carboxylic Acid	R-C(=O) – OH
Anhydride	R-C(=O)–O –C(=O) – R
Ester	R-C(=O)–OR

The C = O stretching frequencies of aldehydes in which R is an alkane are observed at 1735-1710 cm^{-1}. Although the position of the carbonyl absorption does not allow distinction between aldehydes and ketones, the former are recognizable by the C-H stretching vibration that appears as two bands in the 2850-2700 cm^{-1} region.

Double bonds in conjugation with the carbonyl group lower the wavenumber by 25 to 50 cm^{-1}. Thus, aromatic aldehydes (benzene-C(=O)H) generally absorb at 1700-1690 cm^{-1} and diaromatic ketones at 1670-1660 cm^{-1}. Intramolecular hydrogen bonding to the carbonyl oxygen also lowers the wavenumber by 25 to 50 cm^{-1}.

Carboxylic Acids. Carboxylic acids contain the C(=O)OH functional group. The carbonyl group of acids in which R is not aromatic appears at 1730-1700 cm^{-1} and is shifted to 1720-1680 cm^{-1} when the C=O double bond is conjugated with a C=C double bond.

The most characteristic absorption of carboxylic acids is a broad band from 3300-2500 cm^{-1} due to hydrogen bonded O-H stretching. The C-H stretching vibrations appear as small bands on top of this band.

Amines. Amines are compounds containing the N atom singly bonded to R or H. Primary amines have formula NRH_2, while secondary and tertiary amines have formula NR_2H and NR_3 respectively. Primary and secondary amines show N-H stretching vibrations in the 3500-3300 cm^{-1} region. Primary amines generally have two bands approximately 70 cm^{-1} apart due to asymmetric and symmetric stretching modes. Secondary amines show only one band.

Inter- or intramolecular hydrogen bonding broadens the absorptions and lowers the frequency. In general the intensities of N-H bands are less than of O-H bands. The N-H bending and C-N stretching absorptions are not as strong as the corresponding alcohol bands and occur at approximately 100 cm^{-1} higher frequencies. NH_2 groups give an additional broad band at 900-700 cm^{-1} due to bending.

Nitriles. Nitriles are compounds containing the C(3)N group. A sharp, unusually strong absorption at 2260-2220

cm^{-1} due to CN stretching is characteristic of nitriles. To complete the section on IR spectroscopy, we make one broad and extremely useful generalization.

Every compound, organic or inorganic, has a unique infrared spectrum. This spectrum serves as a fingerprint for the compound. Thus if a compound prepared in the laboratory is thought to be a particular known substance, the fastest way to determine whether or not it is is to obtain the IR spectra of the compound and the known substance. If the spectra are identical (band-for-band match), then so are the substances.

ELECTRON MOTION IN MOLECULES—EELECTRONS

The electrons can change energy states within the molecule. When they do this, they change location with respect to the atomic nuclei. Motions of electrons within the molecule are associated with frequencies that are typically in the range 10^{14} to 10^{15} s^{-1}.

When this frequency matches the frequency of light impinging on the molecule, the light can be absorbed, and an electron moves to a higher energy state.

The frequencies of light absorbed give us information about the motions of the electrons in the molecule. Photons with frequency 10^{15} s^{-1} have energy of 6.6×10^{-19} J, and fall in the visible and ultraviolet regions of the electromagnetic spectrum. Spectroscopy involving changes in electron motions is therefore commonly called UV-Visible spectroscopy. It is also called electronic absorption *spectroscopy* (EAS).

The wavelengths of light absorbed and the extent of absorption by a substance can be measured using an instrument called an electronic absorption spectrometer. A typical spectrometer covers the wavelength range between 1100 and 190 nm and thus ranges from the near IR region (1100-750 nm) to the near UV region (400-190 nm) ("near" means bordering on the visible region). To measure the UV-Vis spectrum of a substance requires the preparation of a solution of the substance in an appropriate solvent.

The spectrum is a plot produced by the spectrometer of the amount of light absorbed (the absorbance) as a function of the wavelength of the light.

The spectrum of potassium permanganate, $KMnO_4$, dissolved in water. Note that the absorption does not consist of sharp lines, as in NMR, or even as relatively narrow signals, as in IR. Instead, absorption usually occurs over a range of wavelengths. Each absorption region is therefore called an *absorption band*, to indicate that it covers a band, or range, of wavelengths, rather than occurring at just a single wavelength.

The amount of light absorbed by the substance — the absorbance — at a particular wavelength is directly proportional to 1) the length of solution through which the light beam passes; and 2) the concentration of substance (number of molecules or moles per unit volume) in the solution.

$$A = e\,l\,M$$

where A = absorbance (amount of light absorbed); l = the length of solution through which the light passes, in cm; M = the molarity (number of moles of substance per liter) of the solution; and e = the proportionality constant relating A to l and M. e is called the molar absorptivity of the substance at the particular wavelength at which A is measured. The most important aspect of this equation for us at this point is this: the amount of light absorbed is directly proportional to the amount of substance per unit volume of solution. This is called Beer's Law.

The molar absorptivity is characteristic of the substance and the wavelength. Values of e for a substance are usually quoted for the wavelengths of maximum absorbance, l_{max} — that is, the tops of the absorption bands. The value of e gives a measure of the effectiveness of a molecule of substance at capturing a photon of light. Values can range from as low as 0.01 to as high as 10^5 L/mole-cm. Note that substitution of 1000 cm^3 for 1 L reduces the units of e to cm^2/mole.

This shows that e measures the effective cross-sectional area of the substance with respect to capturing light. Since the e values of a substance are characteristic of it, we may use the amount of light absorbed to measure the concentration of the substance in solution.

Example. The spectrum of a mixture of $KMnO_4$ and $K_2Cr_2O_7$ dissolved in water has A = 0.266 at 565 nm and A = 0.790 at 350 nm. For $KMnO_4$, e = 1.13×10^3 at l = 350 nm and 1.27×10^3 at 565 nm. For $K_2Cr_2O_7$, e = 2.84×10^3 at l = 350 nm and essentially zero at 565 nm. Calculate the concentrations of $KMnO_4$ and $K_2Cr_2O_7$ in the solution.

Solution. Since dichromate does not absorb at 565 nm, the absorbance at 565 is due entirely to permanganate. It's concentration can be calculated from Beer's Law:

$$[MnO_4^-] = \text{A at 565/e at 565} = 0.266/1.27\times10^3$$
$$= 2.09\times10^{-4}\ M$$

The absorbance at 350 nm is a sum of contributions from the two species; however, the permanganate contribution can be calculated and subtracted, allowing the dichromate concentration to be obtained:

$$A_{MnO4-}\text{ at }350 = (\text{e at }350)\times[MnO_4^-]$$
$$= 1.13\times10^3\times2.09\times10^{-4}$$
$$= 0.236$$
$$A_{Cr2O72-}\text{ at }350 = A_{total}-A_{MnO4-}$$
$$= 0.790-0.236 = 0.554$$
$$[Cr_2O_7^{2-}] = (\text{A at }350)/(\text{e at }350)$$
$$= 0.554/2.84\times10^3$$
$$= 1.95\times10^{-4}\ M$$

Example. The complex ion, $FeL(CH_3CN)(CO)^{2+}$ has the structure shown in Figure. L represents the cyclic structure consisting of carbon, hydrogen, and nitrogen atoms. The complex decomposes when dissolved in acetonitrile, CH_3CN, according to the following equation

$$FeL(CH_3CN)(CO)^{2+} + CH_3CN \rightarrow FeL(CH_3CN)_2^{2+} + CO(g)$$

The course of this reaction can be monitored by measuring the absorbance of the solution at 556 nm, a wavelength where $FeL(CH_3CN)_2^{2+}$ absorbs but $FeL(CH_3CN)(CO)^{2+}$ does not. The following absorbance data are obtained at various times after the reaction begins:

Time, minutes	*Absorbance*
0	0
3	.121
8.5	.243
11.5	.302
14.5	.369
17	.397
20	.447
24	.487
28	.528
36.5	.579
45	.615
59.5	.646

The initial concentration of $FeL(CH_3CN)(CO)$ was 0.694×10^{-4} M. The value of e for $FeL(CH_3CN)_2^{2+}$ at 556 nm is 9650 $M^{-1}cm^{-1}$.

- Plot $[FeL(CH_3CN)(CO)^{2+}]$ vs time.
- What percentage of the reactant remains after 45 minutes?
- What is the average rate of the reaction over the first 15 minutes?

Solution. The concentration of $FeL(CH_3CN)(CO)^{2+}$ at each time must be calculated from the measured absorbance, the molar absorptivity of the product, $FeL(CH_3CN)_2^{2+}$, and the known initial concentration of $FeL(CH_3CN)(CO)^{2+}$. For example, at t = 3 minutes, the measured absorbance is 0.121. The corresponding $[FeL(CH_3CN)_2^{2+}]$ is

$$[FeL(CH_3CN)_2^{2+}] = A/e = 0.121/9650 = 1.25\times10^{-5}\ M.$$

The reaction stoichiometry indicates that 1 mole of product is obtained from 1 mole of reactant. It follows that $[FeL(CH_3CN)(CO)^{2+}] = (6.94-1.25)\times10^{-5} = 5.69\times10^{-5}$ M. An analogous calculation can be performed at each time in the table. This enables a plot of $[FeL(CH_3CN)(CO)^{2+}]$ versus time to be constructed.

Find the concentration of $FeL(CH_3CN)(CO)^{2+}$ from the curve at t = 45 minutes. Divide this by the initial concentration of reactant to find the fraction remaining at 45 minutes. Then multiply by 100.

When t = 45 min, the curve indicates that

$[FeL(CH_3CN)(CO)^{2+}]$ has fallen to 5.7×10^{-6} M. The fraction of reactant remaining is thus

$5.7\times10^{-6}/6.94\times10^{-5} = 0.082$, or 8.2 per cent

Divide the change in $[FeL(CH_3CN)(CO)^{2+}]$ that occurs over the first 15 minutes by the elapsed time, 15 minutes. This gives the average change in concentration per minute over this time interval:

$(6.94 - 3.05)\times10^{-5}$ M/15 min = 2.59×10^{-6} M/min

Example. The electronic absorption spectrum of a water solution of $CrCl_3$ (chromium(III) chloride) is shown in Figure. The absorption bands, resulting from the Cr^{3+} ion, are in the visible region of the spectrum, and are responsible for the violet colour of the solution. What are the energies of the most strongly absorbed photons? Plot the electronic energy levels of the Cr^{3+} ion on an energy level diagram.

Solution. The most strongly absorbed photons have energies corresponding to the wavelengths at the tops of the absorption bands. The energies, read directly from the absorption maxima in the spectrum, are at 16500 and 23000 cm^{-1}. These are in units of reciprocal wavelength and must be converted to energy units using the relationship

E = hn = hc/lFor the 16500 cm^{-1} photon, E = $(6.626\times10^{-34}$ J/s$)(3.0\times10^{8}$ m/s)(16500 cm^{-1})(100 cm/m) = 3.28×10^{-19} JFor the 23000 cm^{-1} photon, E = 4.57×10^{-19} J, by the same method.

To construct the energy level diagram, we must realise that the observed absorptions correspond to transitions of an electron in the Cr^{3+} ion from the lowest energy state to two higher energy states; one corresponding to each absorption band. The photon energies calculated above give us the gaps between the energy levels. The energy level diagram should thus show three states. The middle state should be 3.28×10^{-19}J above the lowest one. The highest state should be 4.57×10^{-19} J above the lowest one. The diagram is shown in Figure.

The Correspondence of Colour and Spectrum.

Many substances are colored, resulting from absorption

of light in the visible region of the spectrum. In addition to making these substances visually interesting, colour is an important physical property, and is used with such quantities as melting point and boiling point in describing the substance. It is important to have a qualitative understanding of the relationship between the colour of light absorbed by a substance and the colour of light that we perceive when we look at the substance.

These colors are *complementary.* We see the light that is NOT absorbed by the colored substance. The colour that we see is that of the light that is not absorbed. Although this statement may contradict what you have previously thought, it nonetheless makes sense. Photons that are absorbed by a colored substance cannot reach the eye to be observed. The photons reaching the eye are those that are NOT absorbed.

White light is made up of the three primary colors, red (R), green (G), and blue (B). The relationships between and combinations of these three colors are best shown in the form of the colour wheel. The wheel is set up with the primary colors at 12, 4, and 8 of a clock face. The complements of the primaries are shown directly opposite the corresponding primary. Thus cyan, the complement of red, is at 6 oclock. Cyan, C, is a combination of equal amounts of B and G, which occur to either side of it.

Similarly, magenta (R + B) occurs opposite its complement, green; and yellow (R + G) appears opposite its compement, blue. A substance that absorbs light of a particular colour appears to have the colour opposite the absorbed colour on the wheel.

Thus, if red is absorbed, cyan is seen. Figure shows the correspondence between colour and wavelength of light, which may be seen by viewing white light through a simple device called a *spectroscope*. The colour of a substance may be understood in terms of its electronic absorption spectrum. Thus if a substance has an absorption band at 550 nm, in the green region of the spectrum, it absorbs green photons, transmits red and blue, and should appear to us to be magenta (R + B).

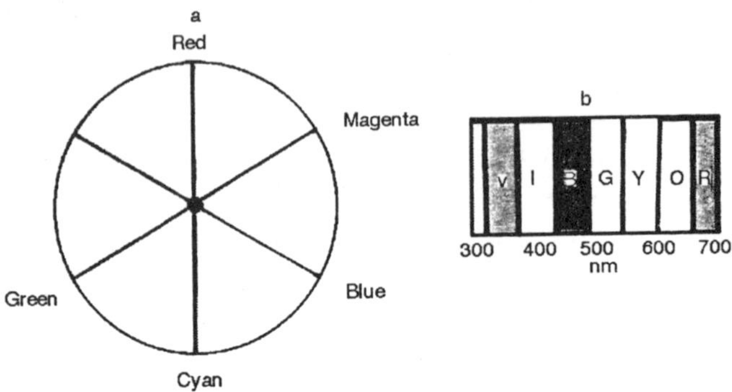

Fig. Colour

Relationship of Structure to the Electronic Absorption Spectrum

It is important to have some idea of the types of molecular electronic structural features that are likely to give rise to absorption bands in the near-IR (1100-750 nm), visible (750-400 nm), near-UV (400-190 nm), and far UV (shorter wavelength than 190 nm) regions of the electromagnetic spectrum. We will make some broad statements for which you can expect many exceptions.

The purpose of the statements is to enable you to make reasonable predictions about electronic transitions in molecules. In making predictions, of course, we run the risk of being wrong. But an incorrect prediction based on sound reasoning is far preferable to no prediction (and no reasoning) at all. You are thus encouraged to be fearless in making predictions.

The near-IR and Visible regions (1100-400 nm). Several structural features are notable for generating light absorption in these regions. First, compounds containing *transition metal ions* (ions of the metals found in the d block of the periodic table) show an interesting rainbow of colors.

These ions are known for their ability to form *complexes*, compounds in which up to six molecules or ions are covalently bonded to the transition metal ion. In this context, the molecules or ions are called *ligands*. In theory, any molecule or ion with at least one lone (nonbonding) electron pair can serve as a ligand by using its electron pair to bond to a transition metal ion.

In practice, molecules containing oxygen (2 lone pairs), nitrogen (1 lone pair), sulfur (2 lone pairs), phosphorus (1 lone pair), or halogen (3 lone pairs) frequently serve as ligands. Table gives formulas and colors for a collection of complexes containing the Co^{3+} ion.

Table. Complexes of Co^{3+}

Complex	*Color*
$[Co(NH_3)_6]Cl_3$	yellow
$[Co(NH_3)_5Cl]Cl_2$	purple
$[Co(NH_3)_4Cl_2]Cl$	green
$[Co(NH_3)_4Cl_2]Cl$	violet

Without going into too much detail, the colors of transition metal ions are a consequence of the partial filling of the d valence orbitals with electrons. When ligands bind to the transition metal, the d orbitals are caused to have different energies, so that the d electrons can undergo light-promoted transitions between them.

The energy spacing between the d orbitals is influenced by the chemical identity of the ligands; thus different ligands coax different colors from the same metal ion. These so-called *d-d transitions* are typically quite weak, with molar absorptivities in the range 0.1-100 $M^{-1}cm^{-1}$

Second, organic molecules with extended *conjugation* (alternating single and double bonds) often absorb visible light.

Acid-base indicators have structures of this type. Extended conjugation normally leads to quite intense colour (strong absorbance, large molar absorptivity), in contrast to transition metal complexes, whose colors are relatively pale. The *porphyrin ring*, which occurs in many biologically-important molecules, has benzene-like resonance involving its double bonds.

It absorbs intensely in the visible region. Electronic absorption spectral data for a number of compounds of this type are presented in Table.

Table: Spectral Data for Conjugated Species

Molecule	*Absorption Wavelengths and Molar Absorptivities, $\lambda_{max}(\varepsilon)$*
2-nitrophenol, $C_6H_4(OH)(NO_2)$	279 (6600) 351 (3200)
2-nitroaniline, $C_6H_4(NH_2)(NO_2$	283 (5400) 412 (4500)
Acetophenone, $(C_6H_5)(CH_3C{=}O$	240 (13000) 278 (1110 319 (50)

Third, species (molecules or ions) in which an atom in a high oxidation state is bonded to species with lone pairs of electrons (such as oxide ions and halide ions) are often colored, due to the transition of an electron from one of the lone pairs to the electron-deficient high-oxidation-state atom. Such transitions are called *charge transfers*, because they essentially transfer negative charge (an electron) from a region of negative to a region of positive charge. The permanganate ion, MnO_4^-, is a good example of a species whose colour is due to a charge transfer transition. The anion is intensely purple both in the solid state and in aqueous solution. Table gives spectral data for charge transfer absorptions in a number of species.

Table: Spectral Data for Charge Transfer Absorptions

Species	*Absorption Wavelengths*
MnO_4^-	350 (1130) 565 (1270)
$Fe(CN)_5(C_5H_7N_2)^{2-}$	660 (10^4)

Near UV region (400-190 nm). Many organic molecules with double bonds, particularly those in which one of the double bonded atoms has at least one lone pair, absorb radiation in the near UV region.

Examples include acetone, with a carbon-oxygen double bond and 2 lone pairs on oxygen; benzene, with 3 conjugated double bonds; acetonitrile, with a carbon-nitrogen triple bond and a lone pair on nitrogen; and SO_4^{2-} (an inorganic ion), with sulfur-oxygen double bonds and lone pairs on oxygen.

Table presents absorption data for a number of common molecules of this type.

Table: Spectral Data for Species Absorbing in the Near-UV Region

Molecule	*Absorption Wavelengths and Molar Absorptivities,* $\lambda_{max}(\varepsilon)$
Acetone $(CH_3)_2C{=}O$	279 (13)
Nitromethane CH_3NO_2	275 (15)
Phenol, C_6H_5OH	210.5 (6200) 270 (1450)
Aniline	230 (8600) 280 (1430)

Far UV region (< 190 nm). Molecules containing strong sigma bonds and no lone pairs tend to absorb in this region. The only promotable electrons are those forming the single bonds, and they must receive a large amount of energy in order to undergo a transition to the lowest energy empty orbital of the molecule.

Thus molecules such as ethane, C_2H_6 and other *hydrocarbons* have no absorptions at wavelengths longer than 190 nm. Strongly single bonded molecules with lone pairs on electronegative atoms also tend to absorb only at very short wavelength (high energy).

Thus water, HF, NH_3, and the like have no absorptions in the visible or near-UV regions of the spectrum. Table shows some typical far-UV data.

Table: Spectral Data for Species Absorbing in the Far UV Region

Molecule	*Absorption Wavelengths and Molar Absorptivities,* $\lambda_{max}(\varepsilon)$
Methanol, CH_3OH	177 (500)
Diethyl ether, $(C_2H_5)_2O$	188 (1995) 171 (3981)
Water	167 (7000)
Ethane, C_2H_6	135

The foregoing discussion should give you an idea of the

importance and utility of spectroscopy. The standpoint of theory, have been elucidated largely via spectroscopic methods.

Supplement 1: ^{13}C Nuclear Magnetic Resonance Spectroscopy

In the early days of NMR (the 1950s and 60s), the proton was the only nucleus that could be readily observed in the NMR experiment. It was every carbon chemist's dream to directly observe carbon atoms, but the major isotope of carbon, ^{12}C, has no nuclear spin.

The ^{13}C nucleus has spin 1/2, like the proton, so was in principle observable by NMR. The difficulty is that only one carbon atom in 100 is carbon-13, meaning that the number of nuclei absorbing radio-frequency photons would be very small. Early on, the absorptions were indeed too weak to be observed. In the intervening years, however, tremendous improvements in NMR hardware and methods have been made, so that now the chemist can routinely obtain the NMR spectrum due to ^{13}C atoms in a sample.

The characteristics of ^{13}C NMR spectra are fundamentally similar to those of proton spectra. That is, a signal is obtained for each structurally different type of carbon atom; the positions of the signals are measured in ppm with respect to the ^{13}C signal for TMS; the signals tend to be further downfield for C atoms attached to electronegative atoms like Cl and O; and it is possible to observe spin-splitting of the ^{13}C resonances by protons that are directly attached to the observed carbon. However, there are some differences in detail between ^{1}H and ^{13}C spectra. First, the chemical shift range exhibited by carbon is more than 10 times larger than that of protons.

Thus although most proton resonances occur within 10 ppm downfield from the TMS resonance, carbon resonances can range more than 100 ppm from the TMS ^{13}C signal. Second, although there is still one signal from each different type of carbon atom, the intensities of the signals do not provide information about the number of carbon atoms generating the signal. Recall that this was one of the very useful aspects of proton NMR; it is not useful for carbon. The reasons for this are complex and we need not bother with them. Third, we expect the signal from a particular type of carbon atom to be split into subsignals by protons attached to the carbons.

Indeed such splitting can be observed, and it follows the usual rules, but most often chemists choose NOT to observe it. Instead, they set the NMR instrument to electronically *decouple* the proton spins, so that only a single line is seen for each chemically distinct type of carbon atom. Fourth, ^{13}C spectra are often obtained by dissolving the compound of interest in *deuterated chloroform*, $CDCl_3$. The carbon atom of the solvent gives rise to an NMR signal that is split into three subsignals due to coupling with the deuterium (heavy hydrogen) atom. This triplet of signals always falls at about 77 ppm downfield from the TMS signal. It should be ignored when interpreting ^{13}C NMR spectra.

The ^{13}C NMR spectrum for chloropentane. The spectrum has the following features:

- It shows a TMS signal at 0 ppm, a solvent signal at 77 ppm, and 4 additional signals, all singlets.
- The signals all fall within 80 ppm of the TMS signal.
- The 4 singlets have different intensities (heights).

The spectrum tells us immediately that there are 4 distinct types of carbon atom in the molecule, although it does not tell us how many of each. Figure shows several possible structures (isomers) for chloropentane.

Of those shown the observed spectrum could only arise from options D, E, and G, all of which have 4 carbon atoms in structurally different environments. Without further information, we cannot say which isomer we have. However, suppose that we then re-run the spectrum, this time allowing the attached protons to couple and split the ^{13}C signals, to obtain the result in Figure..

The small letters in the spectrum indicate the extent of splitting of each signal: s for singlet, d for doublet, and so on. Of structures D, E, and G, only D is consistent with the observed proton couplings. It should produce quartets from carbons 1 and 4, each coupling with 3 protons; a triplet from carbon 2 coupling with 2 protons, and a singlet from carbon 3 coupling with no protons.

Structure E is expected to produce quartets from carbons 1 and 4 and doublets from carbons 2 and 3, while G should show a quartet from carbons 1, 2 triplets from carbons 3 and 4, and a doublet from carbon 2. Let's look at a couple more examples.

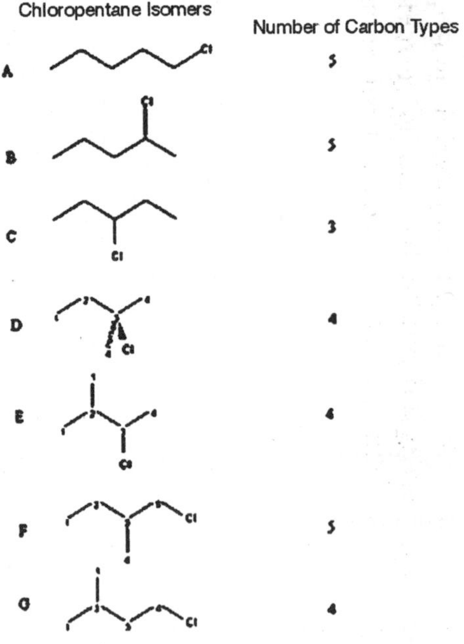

Fig. Structural Isomers of Chloropentane

Example. The structure and proton NMR of isobutyraldehyde are shown in Figures, respectively. Predict the appearance of the ^{13}C NMR spectrum of this molecule.

Solution. The structure shows 3 types of carbon atom, so 3 signals are expected. If coupling to protons is allowed, the signal from the CH_3 carbons should be a quartet, that from the carbonyl carbon should be a singlet, and the final carbon should produce a doublet. The signal from the carbonyl carbon should be quite far downfield due to the proximity of oxygen.

Example. A compound of formula C_6H_{12} produces the ^{13}C spectrum. What is the molecular structure of the compound?

Solution. Possible atom arrangements corresponding to the molecular formula are shown in Figure. Clearly there are quite a few of them! The structures are expected to produce the following numbers of ^{13}C signals:

A — 6
B — 6
C — 3
D — 6
E — 6
F — 5
G — 5
H — 5
I — 6
J — 5
K — 2
L — 1
M — 4

Because 4 signals are observed in the spectrum the structure must be M, which is called methylcyclopentane. If coupling with attached hydrogen atoms were allowed, we would expect to see a quartet, a doublet, and 2 triplets. You might want to predict the splitting pattern that would be observed for each of the other possible structures.

It may have occurred to you to wonder whether coupling of one ^{13}C nucleus with another can occur, and whether it is observable in the NMR spectra. In fact, it is not observable; perhaps we should be thankful for this, as it would greatly complicate things! We should spend a bit of time to understand why this type of coupling is not seen.

As said above, the natural abundance of ^{13}C is about 1 per cent (actually 1.11 per cent, but we use the simpler number because calculations are easier). This means that 1 carbon atom in a hundred has mass number 13. We can use this fact to determine the probabilities that varying numbers of ^{13}C atoms will be found in a molecule with, say, 4 carbon atoms. Let's begin with the probability that a particular molecule will have NO ^{13}C atoms.

Here is the question we are asking: "What is the probability that carbon 1 is ^{12}C and that carbon 2 is ^{12}C and that carbon 3 is ^{12}C and that carbon 4 is ^{12}C?" In probability mathematics, and means multiply the probabilities for the events that it links. Thus the probability that we seek is $(99/100)\times(99/100)\times(99/100)\times(99/100) = (0.99)^4 = 0.96$. So of 100 of these molecules, 96 will have no carbon-13 atoms in them! The remaining 4 will have between 1 and 4 carbon 13 atoms. Next, let's calculate the probability that a molecule contains four carbon-13 atoms by calculating $(1/100)^4 = 1\times10^{-8}$.

This is one atom in 100 million, such a small number that we can ignore this possibility. Much more likely is that a molecule will contain a single carbon-13 atom; this situation can be shown to have probability of about 0.039.

The probability that a molecule will contain 2 carbon-13 atoms next to each other, a situation required for spin-spin coupling, is $2.9'10^{-4}$. Only 3 molecules in 10000 qualify! Thus the intensity of the NMR signal from the molecules containing one carbon-13 atom is expected to be more than 100 times greater than the intensity from molecules with two neighboring carbon-13 atoms; the latter is so weak that it will not be seen. Only coupling with attached protons is significant in carbon NMR.

The modern chemist, armed with the techniques of NMR, IR, and UV-visible spectroscopy, is able to discern the detailed structure of just about any molecule of interest. Even large biological molecules yield to some of these techniques. We can be confident that in years to come, science will develop yet other ways to exploit the interaction of light and matter to reveal nature's details at the molecular level.

Supplement 2: An Intuitive Development of Beer's Law

It is possible to arrive at Beer's Law using a simple macroscopic physical system consisting of an inclined plane perforated with holes at a certain density, and divided into four sections of equal length. The plane is pictured in Figure.

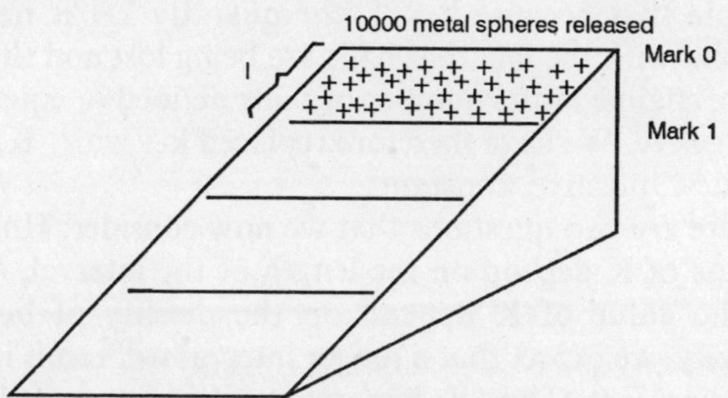

Fig. Beer's Law Inclined Plane

Suppose that 10000 identical small steel spheres are released at the top of the plane, and that we observe that 9000 arrive at the end of the top section, having avoided falling into one of the holes.

Based on this information, how many spheres will arrive at the bottom of the plane? Two approaches seem reasonable on the surface. The first is to assume that the same number of spheres will fall into holes in each section. If 1000 fell in the first section, then a total of 4000 will fall along the entire length of the plane, leaving 6000 to reach the bottom.

The second approach is to assume that the number of spheres that fall into holes in a given section is proportional to the number that enter the section; i.e., that the same fraction of spheres will disappear in each section. By this reasoning, we predict that 8100 spheres will survive the first 2 sections, 7290 will survive through sections, and 6561 will reach the bottom. This second line of reasoning proves correct: the number of balls that survive a section is proportional to the number of balls that start into the section. We write this in equation form: $N_n = k \times N_{n-1}$

This equation states that the number of spheres arriving at mark n is proportional to the number of spheres that leave the preceding mark, n-1. k is a proportionality constant that must have a value between 0 and 1; it has the value 0.9 for our example above. We can also express the equation in terms of the change in the number of balls:

$$DN = N_n - N_{n-1} = k \times N_{n-1} - N_{n-1} = N_{n-1} \times (k-1) = -K \times N_{n-1}$$

Note that because $k < 1$, the quantity k-1 is negative, consistent with the fact that balls are being lost and therefore that the change in the number of balls defined in equation is also negative. We have therefore replaced k-1 with –K, where K is a new, positive, constant.

Here are two questions that we now consider: How does the value of K depend on the length of the interval, *l*? How does the value of K depend on the density of holes, r? Intuitively, we expect that a longer interval will result in more balls being lost. Thus K increases as *l* increases. Also, we expect more balls to be lost if the density of holes is increased. So K increases as r increases. In equation form, this becomes:

$$K = \zeta \times l \times r$$

where z is a constant representing the number of balls lost per unit density of spheres per unit length of section. Substituting 4S-3 into 4S-2 gives

$$: \Delta N = -\zeta \times l \times \rho \times N$$

Assuming an interval of infinitesimal length will result in an infinitesimal loss of spheres:

$$: dN = -z \times r \times N \times dl$$

Separating the variable, N, from the variable, *l*, gives

$$: dN/N = -\zeta \times \rho \times dl$$

Finally, integration provides

$$: \ln(N/N_o) = -\zeta \times \rho \times l$$

Translating to base-10 logarithms gives:

$$\log(N/N_o) = -(1/2.303) \times \zeta \times \rho \times l = -e \times \rho \times l$$

Of course this equation can be expressed in exponential form as $N/N_o = 10^{-e \times r \times l}$. In words, this equation states that the fraction of balls surviving an interval decreases exponentially with the length of the interval and the density of holes in the interval.

Let's now make a direct analogy between our inclined plane and a solution of a substance that absorbs light at some wavelength.

We think of *l* as the length of solution through which the light beam must pass; r as the density (concentration) of absorbing molecules in the solution; the small spheres as photons of light; and N/N_o as the fraction of light transmitted by the solution:

$$\log(I/I_o) = -e \times M \times l$$

We have replaced the density of holes with the concentration of the solution in moles absorbing substance per liter, and the number of spheres with the intensity, I, of light. The intensity is simply the number of photons per second arriving at a particular location, so is a close analog of N. We now define a quantity called the absorbance, A, as $A = e \times M \times l$ so that equation becomes:

$$\log(I_o/I) = A.$$

A represents the amount of light absorbed (the number of spheres lost). Note that when all light is transmitted by a sample, $I_o/I = 1$ and $A = 0$. When 1/10 of the light is transmitted, $I_o/I = 10$ and $A = 1$. Thus A goes up as the fraction transmitted goes down. The equation $A = e \times M \times l$ is, of course, Beer's Law.

Chapter 5

The Gas Phase

The phases (physical states) in which a pure substance may exist. For most substances there are three phases — solid (s), liquid (l), and gas (g). The gas phase is the simplest of the three to deal with theoretically, primarily because the molecules behave completely chaotically and therefore give rise to simply-formulated average properties. The solid phase is also relatively easily dealt with, because it is so highly organized.

Gases and their properties play an ubiquitous and critical role in our daily lives. Life is supported by the oxygen of the air, which reacts with glucose in biochemical combustion. Living systems have evolved complex molecular architecture to absorb oxygen from the air; carry it to the cells; deliver it to the cells; transport it within the cell; and supervise its multistep, controlled reaction with glucose. It is the rapid expansion of gases produced in the explosive combustion of gasoline that performs the work of the internal combustion engine. Expansion and contraction of gases in response to temperature and pressure changes is at the basis of continually changing weather patterns.

Clearly an understanding of the properties, behaviour, and chemical reactivity of gases is an important foundation for developing an appreciation of complex biological and technological systems. Following a discussion of the ideal gas law, a simple mathematical model for gas behaviour, we discuss the Kinetic Molecular Theory of gases, which establishes a link between molecular motion and temperature.

MACROSCOPIC PROPERTIES OF GASES.

All gases have observable properties in common:

- Gases flow readily from one space to another and occupy all available space.
- Gases assume the shapes of their containers.
- Gases are readily compressible. By exerting force, we can cause a gas to occupy a smaller volume.
- Two or more gases form homogeneous mixtures (solutions) in all proportions. An example is air, a mixture of primarily nitrogen and oxygen with small amounts of other gases.
- Gases diffuse rapidly. This means that a gas can move across a large space in a relatively short time. Examples are the odors of perfume and skunk.

The state of a gas is described by the values of 4 macroscopic variables: the volume (V) occupied by the gas; the temperature (T) of the gas; the pressure (P) that the gas exerts on the walls of its container; and the amount of gas in moles (n). The values of these quantities are related by a very simple equation called the ideal gas law. We will discuss this equation after a brief word on pressure.

Pressure.

Pressure is defined as force per unit area. We illustrate the concept of pressure with an example.

Example. The bottom of each foot of a 150-pound man has a surface area of about 24 square inches. When the man stands on a bathroom scale, what pressure is exerted by his feet?

Solution. The force exerted by the feet is the man's weight, 150 lb. This is distributed over the area of the bottoms of both feet, or 48 square inches. The pressure is then

$$P = F/A = 150 \text{ lb}/ 48 \text{ in}^2 = 3.1 \text{ lb/in}^2$$

If the man stands on the scale with only one foot, the force is the same, but the pressure is doubled because the area is cut in half.

Gases exert pressure on the walls of their containers. Since a gas uniformly occupies a container, it exerts the same pressure on all the walls. The gases that make up the atmosphere exert pressure on the surface of the earth. This pressure, called the atmospheric pressure, can be measured using a barometer. The left side of the figure shows a beaker or cup containing mercury. A glass tube of length 1 m and cross-sectional area, A, fitted with a stopcock on top, is inserted into the beaker so that the end of the tube is below the level of mercury in the beaker. Initially, the stopcock is open and the levels of mercury inside and outside the glass tube are the same. A vacuum pump is then attached to the top of the glass tube, and the air is pumped out. As this occurs, the mercury level rises in the tube until it reaches a height, h.

Despite further pumping, the mercury level rises no higher in the tube. When all air has been pumped out of the tube, the stopcock is closed, and the mercury maintains its position in the tube. This behaviour is understood as follows. Initially, before pumping, the atmosphere exerts the same pressure on the surfaces of mercury inside and outside the tube. Since $P_{out} = P_{in}$ = atmospheric pressure, the mercury levels inside and outside the tube are the same. As air is pumped out of the tube, its pressure on the mercury inside the tube decreases, and mercury rises in the tube to a height, h, such that the force due to the mass of mercury in the tube just counterbalances the pressure of the atmosphere on the mercury outside the tube.

The height h is found to vary from day to day because the pressure of the atmosphere varies with weather conditions. However, its value is always in the neighborhood of 0.76 meters. For this reason, a unit of pressure called the atmosphere (atm) is defined as the pressure that will support a column of mercury 76 cm, or 760 mm, in height.

$$1 \text{ atm} = 760 \text{ mm Hg}$$

The pressure exerted by a column of mercury 1 mm in height is called a torr after Evangelista Torricelli, inventor of the barometer. Thus

$$1 \text{ atm} = 760 \text{ mm Hg} = 760 \text{ torr}$$

In the English and SI systems, the units of pressure are respectively the pound per square inch and the pascal (1 Pascal = 1 Newton per square meter, where a Newton is the force required to accelerate a 1 kg body at 1 m/s^2). The pressure in Pa exerted by a column of mercury 760 mm in height can be readily calculated and will provide a conversion factor between SI and traditional non-SI units.

The force exerted by a mass, m, of mercury is mg, where g is the acceleration due to gravity. This force is distributed over the cross sectional area, A, of the tube. The mercury therefore exerts the pressure, P = mg/A on the surface of mercury at the bottom of the tube. The mass of mercury can be calculated from its density, r, and its volume, which is Ah. Substituting for m in the expression for pressure gives equation:

$$P = \rho \times (Ah)g/A = \rho \times gh$$

The pressure exerted by a column of mercury (or any liquid) of height h is independent of the cross sectional area of the column. The diameter of the barometer tube is therefore unimportant. Substituting the density of mercury (13.595 g/cm^3), the acceleration constant (g = 9.80665 m/s^2), and h = 0.760 m in 5–1–1 gives P = 1.0132 × 10^5 Nm^{-2} (1 Nm^{-2} = 1 Pascal). The following conversion relations are thus obtained:

1 atm = 14.7 lb/in^2 = 1.013 × 10^5 Pa (1 Pa = 1N/m^2)
= 1.013 bar (1 bar = 10^5 Pa)

Although the bar is official, the atmosphere and the torr are still the most commonly used pressure units in chemistry. We will therefore use the atm and the torr.

Note that atmospheric pressure is 14.7/3.1 = 4.7 times greater than the pressure exerted by the feet of the man in Example. The magnitude of the pressure exerted by the atmosphere is equivalent to having that 150-lb man, holding a 550-lb weight, standing on your stomach.

The pressure of a gas confined in a container cannot be measured with a barometer. For this a variety of devices may be used, the simplest being the manometer. A manometer is a U-shaped glass tube containing mercury or some less toxic fluid. The gas container is attached to one side of the manometer, and the other side is either open to the atmosphere (an open-end manometer) or evacuated (a closed-end manometer).

The difference in the levels of fluid in the two arms of the manometer is a measure of the pressure of gas in the bulb through the relationship, $P_1 = P_2 + P$. Here P_1 is the pressure exerted by the gas on arm 1 of the manometer, P_2 is the pressure exerted on the arm of the manometer not connected to the gas container (arm 2), and P is the height of fluid in arm 2 minus the height in arm 1. P is positive if fluid is higher in arm 2 and negative if it is higher in arm 1.

For an open-end manometer, P_2 is atmospheric pressure; for a closed-end manometer, P_2 is zero. If the manometer fluid is mercury, the pressure is in torr and may be converted to atm using the conversion factor given above. If the manometer contains some other fluid, the height of fluid may be converted to an equivalent height of mercury using the densities of the fluid and mercury.

Example. An open-end manometer containing water as the working fluid is used to measure the pressure of gas in a bulb. The level of water in the side attached to the bulb is 8.65 cm higher than that in the side open to the atmosphere. Atmospheric pressure is 693 torr. What is the pressure of gas in the bulb?

Solution. Use the relation $P = r \times g \times h$. To find the height of mercury equivalent to 8.65 cm of water, we equate $r \times gh$ for water to $r \times gh$ for mercury. The gravitational acceleration cancels, giving

$$(1.00 \text{ g/cm}^3)(86.5 \text{ mm}) = (13.60 \text{ g/cm}^3)(h)h(Hg) = 6.4 \text{ mm}$$

The pressure of gas is this amount less than atmospheric pressure:

$$P_{gas} = 693 - 6.4 = 687 \text{ torr.}$$

THE IDEAL GAS LAW.

The ideal gas law is a relationship between the pressure (P), the volume (V), the amount (n), and the temperature (T) of a gas. The ideal gas law is called an equation of state, because it represents all possible states — combinations of P,V and T — in which the amount of gas, n, can be found.

$$PV = nRT$$

In this equation, P is the pressure in atm; V is the volume

in L; n is the amount of gas in moles, and T is the Kelvin temperature. R is a proportionality constant, remarkably the same for all gases, which relates the 4 variables. Equation is of simple form, which belies the effort expended in developing it. Over 200 years elapsed between the first quantitative experiments on gases and the final enunciation of equation. The simple form of the equation hides a remarkable sophistication in understanding. We now examine what the equation says regarding gas behaviour.

Volume-Pressure Relation.

If we rearrange the equation to solve for the volume, the equation states that volume decreases as the pressure exerted on (also by) the gas increases. The inverse relation of pressure and volume is a fact of every day experience. For example, when you compress the air in the chamber of a tire pump, you feel the increase in pressure of the air within.

In 1662, Robert Boyle developed the quantitative relationship between pressure and volume that we now know as Boyle's Law, which states that the volume occupied by a fixed amount of gas at a fixed temperature varies inversely with the pressure exerted on the gas. Boyle's Law was the first-discovered component of equation. Note that the ideal gas law predicts that the volume should fall to 0 as pressure gets very large. For real gases this does not happen.

At ordinary temperatures, most real gases will liquefy if pressure is made large enough. Once liquefication has occured, volume does not decrease much with increasing pressure. Real gases obey equation closely only when P is not too large. This is why equation is called the ideal gas law — it applies strictly only to an idealized gas, which does not liquefy but instead shrinks to zero volume when pressure is made very large.

Volume-Temperature relation.

The ideal gas law states that the volume occupied by an amount of gas, n, varies directly with the absolute temperature of the gas. A particular sample of gas, then, should double in volume if heated from 300 to 600 K. This behaviour, too, is quite well known, at least qualitatively, in every day experience, as in the expansion (contraction) of a heated (cooled) balloon.

The relationship between gas volume and Celsius temperature was first established by Jacques Charles in 1800. He clearly established the linear relationship between the two quantities. Similar experiments relating gas pressure and temperature were carried out at about the same time by Gay-Lussac, and similar proportionality was found. In both cases, the relationships obtained were not so simple as the ideal gas law, because the absolute scale of temperature was not then recognized. The relation between volume and Celsius temperature is in equation.

$$V = aXt(^oC) + b$$

Plots of V versus t(°C) for gas samples containing different amounts of gas give different slopes and different intercepts on the V axis. However, they all give the same intercept on the t axis, -273.16 °C. This suggests that the zero of temperature be shifted 273.16 Celsius degrees to the left. The resulting scale of temperature is, of course, the Kelvin scale. Equation takes the much simpler form, $V = a \times T$.

A profound implication of equation is that there exists a temperature at which the volume of gas should become zero. This temperature is called absolute zero. The Kelvin scale is set up so that 0 K corresponds to the temperature at which the volume of an ideal gas sample would become zero. This corresponds to -273.16 on the Celsius scale. Any temperature lower than this value would cause the gas to have negative volume. Since this is physically meaningless, the conclusion is that temperatures lower than absolute zero do not exist. That there is a lower limit on attainable temperature is not obvious from our everyday experience and from the manner in which we normally measure and think about temperature; the conclusion is nonetheless correct.

Relation Between Volume and Amount of Gas.

The ideal gas law states that the volume occupied should

increase in direct proportion with the amount (moles) of gas, when the gas is kept at the same pressure and temperature throughout. This is an extension of the hypothesis by Amadeo Avogadro in 1811 that equal volumes of different gases contain the same number of molecules. Avogadro made this statement in explanation of many observations about the relative volumes of gaseous elements that react to form compounds. It was in fact this hypothesis that finally allowed correct molecular formulas to be deduced and the atomic mass scale to be put on a firm basis.

The Value of the Gas Constant R.

The gas constant R turns out to be an ubiquitous and extremely important quantity in physical science. We begin by obtaining a value for it based on the experimentally-measured volume of 1.0 mole of gas at 1.00 atm pressure and a temperature of 273.16 K. It is found that under these conditions the gas occupies a volume of 22.414 L. Rearranging equation to solve for R, we can calculate its value:

$$R = PV/nT = (1.000 \text{ atm})(22.414 \text{ L})/(1.000\text{mole})(273.16\text{K}) = 0.08206 \text{ L-atm/K-mole}$$

The numerical value of R depends upon the units in which pressure and volume are expressed. P is commonly expressed in atmospheres or torr (both non-SI) or Pascal (SI), and V is commonly expressed in mL. Values of R in common P-V unit combinations are presented in Table.

Table. In Common Pressure-Volume Unit Systems

Pressure Unit	*Volume Unit*	*R*
Atm	L	0.08206 L-atm/K-mole
Bar	L	0.08314 L – bar/K-mole
Torr	mL	6.237×10^4mL-torr/K-mole
Pa	m^3	8.314 J/K-mole

From the Pascal-m^3 value, we see that R has units of energy/K-mole. It follows that the combined units L-bar, L-atm, and mL-torr are also energy units. We can express R in terms of any desired energy unit by using an appropriate conversion factor.

In using the ideal gas law, equation, the units of P, V,

and R must be self-consistent, and the temperature must be expressed in Kelvins. A few examples of the use of the ideal gas law are now presented.

Example. An ideal gas occupies 326 mL at 730 torr of pressure and a temperature of 19°C. What is its volume at 1.00 atm and 0°C?

Solution. We take two approaches to the solution of this problem. In the first approach, we use the ideal gas law twice. First, we calculate the moles of gas under the original V,T,P conditions; second, we calculate the volume occupied by this number of moles at 1 atm and 0°C.Convert to the necessary units of T, P:

Initial	*Final*
T(K) = 273 + 19 = 292 K	T(K) = 273 K
P = 730 torr = 730/760 atm = 0.961 atm	P = 1.00 atm
V = 326 mL = 0.326 L	V is unknown

Calculate the number of moles of gas under the initial conditions:

$$n = PV/RT$$

$$= (0.961 \text{ atm})(0.326 \text{ L})/(0.08206 \text{ L-atm/K-mole})(292\text{K})$$

$$= 1.307\text{X}10^{-2} \text{ moles}$$

Calculate volume at final T and P:

$$V = nRT/P$$

$$= (1.307\text{X}10^{-2} \text{ moles})(0.08206 \text{ L-atm/K-mole})(273\text{K})/ (1.00 \text{ atm})$$

$$= 0.29 \text{ L (2 significant figures)}$$

To shortcut this long process we take advantage of the constancy of the amount (moles) of gas. Since n is constant, PV/RT must also be constant, and its initial and final values are the same:

$$P_iV_i/T_i = P_fV_f/T_f$$

Solving for V_f and substituting numbers gives V_f = 0.29 L.

Example. 3.24×10^{-4} moles of ideal gas occupies 5.31 mL at 25°C. Calculate the gas pressure.

Solution. Convert units, and apply the ideal gas law to the problem.

Convert units:T(K) = t(°C) + 273 = 25 + 273 =

298KV(L) = 5.31 mL/(1000 mL/L) = 5.31×10^{-3} LCalculate P:P = nRT/V = $(3.24\times10^{-4})(0.08206)(298)/(0.00531)$ = 1.49 atm

Example. Consider the apparatus in Figure. Initially, a quantity of gas is confined in the left box at a pressure of 3.40 atm and a temperature of 286 K. 4 sequential processes, a, b, c, and d, are carried out. Work out the value of each indicated quantity after the completion of each process.

- Valve 1 is opened with valve 2 closed. T is kept constant. What are the final volume and pressure of the gas?
- T is raised to 350 K. What are the final V and P of the gas?
- Valve 1 is closed, valve 2 is opened. What are the final V and P of the gas to the right of valve 1?
- T is lowered to 300 K. What are the final V and P of the gas to the right of valve 1?

Solution. We apply the ideal gas law to each process in turn.

- When valve 1 is opened, the gas will expand to occupy its original box and the 3.0-L box at uniform pressure. T is constant, and the amount of gas is too. The gas volume goes up by a factor of 4, so the pressure goes down by the same factor.
  V occupied by gas = 4.0 LP of gas = 3.4/4 = 0.85 atm
- When T is raised to 350 K, the gas pressure will increase in direct proportion, because V will remain constant. Therefore
  V occupied by gas = 4.0 LP of gas = T_fP_i/T_i = (350)(0.85)/(286) = 1.04 atm
- When valve 1 is closed, the gas in the 1.0-L box will no longer communicate with the gas in the middle box. Only the gas in the middle box will be involved when valve 2 is opened. This gas will expand from its initial pressure of 1.04 atm until its final pressure is 0.25 atm. At this point, the piston in the cylinder will stop moving.

V occupied by gas to right of valve 1 = P_iV_i/P_f = (1.04)(3.0)/(0.25) = 12.48 L.P of gas to the right of valve 1 = 0.25 atm

- When T is lowered to 300 K, the volume occupied by the gas will decrease, as the piston moves in to maintain pressure at 0.25 atm.

  V occupied by gas to the right of valve 1 = V_iT_f/T_i = (12.48 L)(300)/(350) = 10.70 L.P of gas to the right of valve 1 = 0.25 atm

Example. 0.432 g of an ideal gas exerts pressure of 0.46 atm in a volume of 768 mL at 0°C. Calculate the molar mass of the gas.

Solution. The ideal gas law can be expressed directly in terms of MM because moles is mass over MM:

n = PV/RT = (mass)/MM

Rearrange for MM:

MM = mRT/PV, where m is mass.

Solve:

MM =(0.432g)(0.08206 L-atm/K-mole)(273 K)/(0.46 atm)(0.768 L)
= 27.4 g/mole

The gas could be nitrogen, N_2; carbon monoxide, CO; or ethylene, C_2H_4.This method for determining the MM of an unknown gas or volatile liquid is called the Dumas method. It is simple, fast, and reasonable accurate.

Example. Calculate the density of oxygen gas at a pressure of 0.940 atm at room temperature (25°C).

Solution. Rearrangement of the ideal gas law with the recognition that density is mass per volume gives density directly.

Rearrange:

PV = nRT
n/V = m/(MM)V = r/MM = P/RT
ρ = P(MM)/RT

Solve the equation with the specific variables given:

= (0.940 atm)(32.00 g/mole)/(0.08206 L-atm/K-mole)(298K)

$$= 1.23 \text{ g/L}$$

The value has units g/L, and is about 1000 times less than the density of water at the same temperature. Gases are much less dense than liquids, implying that the gas molecules are much further apart than are liquid molecules.

Dalton's Law of Partial Pressure.

The ideal gas law can be applied to mixtures of gases as well as to single pure gases. Consider a mixture of two non-reacting gases, A and B, in a container of volume V. According to the ideal gas law, the pressure exerted by the mixture is given by equation.

$$P_T = n_T RT/V$$

Here n_T, the total moles of gas, is the sum of the moles of A and the moles of B:

$$n_T = n_A + n_B$$

Substitute equation and expand:

$$P_T = (n_A + n_B)RT/V = n_A RT/V + n_B RT/V$$

Suppose that we could remove all of the molecules of B from the container. What pressure would A exert in the container alone? This is easily obtained from the ideal gas law to be $P_A = n_A RT/V$. Similarly, $P_B = n_B RT/V$. Substituting the expressions for P_A and P_B into equation gives:

$$P_T = P_A + P_B$$

This is Dalton's Law of partial pressures. In words, it says that the total pressure of a mixture of gases (P_T) is the sum of the partial pressures of the gases composing the mixture. The partial pressure of gas A is the pressure that n_A moles of A would exert if present alone in the container.

Dividing the expression for P_A by equation gives:

$$P_A/P_T = n_A/n_T = n_A/(n_A + n_B)$$

The ratio of the moles of A to the total moles of all gases in the system is called the mole fraction of A, and is symbolized X_A. Thus $X_A = n_A/n_T$. Substituting this into 5-2-7 and rearranging, we obtain equation:

$$P_A = X_A P_T$$

The partial pressure of A in a mixture of A and other gases is the total pressure multiplied by the mole fraction of A.

Example. A gas mixture containing equal numbers of nitrogen and oxygen molecules exerts 1.32 atm pressure. What is the partial pressure of nitrogen?

Solution. Since half the molecules are nitrogen, half the moles are also nitrogen, so the mole fraction of nitrogen is 0.5.

$$P_{N2} = 0.5\ (1.32\ \text{atm}) = 0.66\ \text{atm}$$

We close this section by emphasizing an implicit assumption of the above treatment. Figure shows two gases, A and B, in a box.

If the box has volume V, what fraction of this volume is occupied by A, and what fraction by B? A frequent answer might be that since half the molecules in the box are A, A occupies half the volume. This is, however, incorrect. If, when we fill the box with gas, we put A in first, the molecules of A will spread out to occupy the whole box. Now add some molecules of B.

Since these molecules are "unaware" of the presence of A, they, too, will spread out in the entire volume. Thus both gases occupy the same volume — the entire volume of the container. $V_A = V_B = V$, the box volume. This is implicit in equations. Similarly, we assumed that temperature is the same for both gases: $T_A = T_B = T$. In contrast, the total moles, n_T, is the sum of the moles of A and B; and the total pressure is the sum of the partial pressures. These relationships are summarized in Table.

Table: Relationships Among State Variables for a Gas Mixture

$$T_A = T_B = T_{mixture}$$
$$n_A + n_B = n_{mixture}$$
$$P_A + P_B = P_{mixture}$$
$$V_A = V_B = V_{mixture}$$

Dalton's Law is particularly useful in experimental situations involving the collection of gases over liquids. In such situations, the space above the liquid contains not only molecules of the gas, but also molecules of the vapour form of the liquid, produced by evaporation of the liquid. The vapour exerts a partial pressure that is part of the total pressure exerted by the gas mixture over the liquid.

This partial pressure due to vapour that exists above the

corresponding liquid phase is the vapour pressure, P_{vap}, of the liquid. According to Dalton's Law,

$$P_T = P_{gas} + (P_{vap})liquid$$

To calculate P_{gas}, we must correct the total pressure for the vapour pressure of the liquid, which we can look up in a table.

Example. 0.1140 g of an unknown metal M reacts with excess aqueous HCl according to the reaction

$$M + 2H^+ \rightarrow M^{2+} + H_2(g).$$

The $H_2(g)$ is collected over water. The volume of the gas above the water is 52.1 mL, the total pressure of the gas is 738 torr, and the temperature is 22.1°C. The vapour pressure of water at 22.1°C is 20.0 torr. Calculate the molar mass of the metal.

Solution. Strategy: $P_T \rightarrow P_{H2} \rightarrow$ moles $H_2 \rightarrow$ moles M $\rightarrow$ molar mass M.Use Dalton's Law to calculate the pressure of H_2 in the gas mixture above the water:

$$P_{total} = P_{H2} + (P_{vap})H_2O$$

$$P_{H2} = P_{total} - (P_{vap})$$

$$H_2O = 738 - 20 = 718 \text{ torr}$$

Calculate the moles of H_2 from the ideal gas law:

n_{H2} = PV/RT = (718/760 atm)(0.0521 L)/(0.08206 L-atm/K-mole)(295.1 K)

$= 2.033 \times 10^{-3}$ moles

Calculate moles M from stoichiometry:

moles M = 2.033×10^{-3} moles $H_2 \times$ 1mole M/1 mole H_2 = 2.033×10^{-3} moles M

Calculate molar mass of M from known mass and moles:

MM = 0.1140g/2.033×10^{-3} moles = 56.1 g/mole

The metal is probably Fe.

Example. Figure shows a syringe that has been filled with $N_2(g)$ to a volume of 20.0 mL. The end of the syringe has been capped with a rubber septum (a membrane that may be penetrated to add or remove substances from the syringe, but which otherwise seals the syringe off). 2.50 mL of pure liquid ethanol, $C_2H_6O(l)$, is injected into the syringe. It is observed that the syringe piston moves out until the occupied volume in the syringe is 32.00 mL. Atmospheric pressure is 730.0 torr, and the temperature of the syringe and its contents is 25.00°C. Calculate the pressure exerted by ethanol vapour in the syringe.Solution. See if you can figure this out on your own.

KINETIC MOLECULAR THEORY.

The ideal gas law was developed from macroscopic observations, with no knowledge of the behaviour of a gas at the molecular level. We now attempt to interpret the ideal gas law in terms of a reasonable molecular model (simple physical picture) of gas behaviour.

What are the individual molecules doing, and why does their behaviour manifest itself in the ideal gas law? We now enter the domain of theory — a mental interpretation of experimental results. The model of gas behaviour currently accepted by scientists is the Kinetic Molecular model. The theory of gas behaviour built on this model is called the Kinetic Molecular Theory (KMT). This theory is one of the oldest and most resoundingly successful in science.

First, we assume that gases consist of very small molecules. Since a gas occupies the entire volume of its container, and since gases readily diffuse, the molecules must move about in space. Consequently they have kinetic energy — energy of motion. Since gases are compressible, the molecules must be far apart. Focussing on an individual molecule, it seems reasonable to assume that it is a small particle that moves about at random, occasionally colliding with a wall of the container and with other gas molecules.

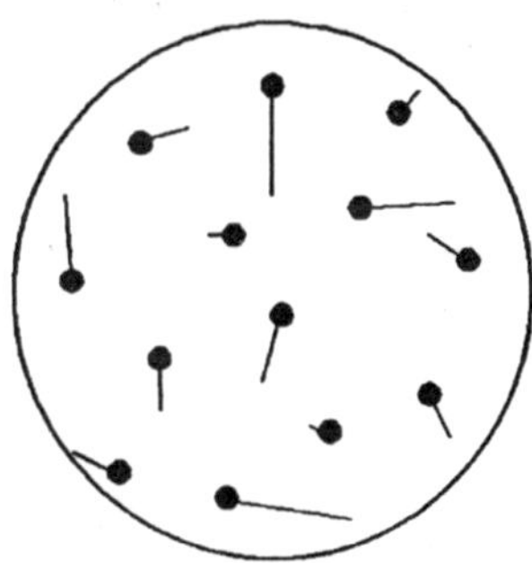

Fig. Dynamic Model of the Gas Phase

The molecule sometimes gains and sometimes loses energy in these collisions, so that it frequently changes speed. Sometimes it moves rapidly, sometimes slowly. Thus our picture of the gas is dynamic rather than static. Figure is an attempt to portray this simple physical model of the gas phase.

These statements about gas molecules are consistent with the properties of gases listed. None of them have been proven, but they seem reasonable and constitute the postulates of the Kinetic Molecular Theory:

- Gases consist of molecules that behave like tiny hard spheres;
- The molecules have no volume — they may be treated as points;
- There are no forces of attraction between molecules;
- Collisions between molecules are elastic (kinetic energy is conserved);
- The molecules are in ceaseless random motion, frequently colliding with the container walls and with each other. A collection of gas molecules is highly disordered.
- The average kinetic energy of the molecules is proportional to Kelvin temperature.

We now pursue the consequences of these postulates. Postulate 6, the origin of which is not obvious. Our aim is to develop an expression for pressure. At the molecular level, pressure must result from collisions of gas molecules with the container walls. A molecule hitting the wall exerts a force on it. The collection of all such forces on a unit area of wall during a given time interval constitutes the pressure. We should be able to calculate P from the force exerted by each collision, multiplied by the number of collisions per unit area of wall:

$$P = \text{Force/Area} = \text{Force/collision} \times \text{Collisions/area}$$

From Newton's second law, force is the rate of change of momentum. Thus

$$F = \text{momentum change/time-collision} \times \text{collisions/area}$$

We move the time factor over to the second term to obtain equation.

F = momentum change/collision×collisions/time-area

To obtain the momentum change per collision, envision a molecule moving directly toward a wall with velocity v, as shown in Figure.

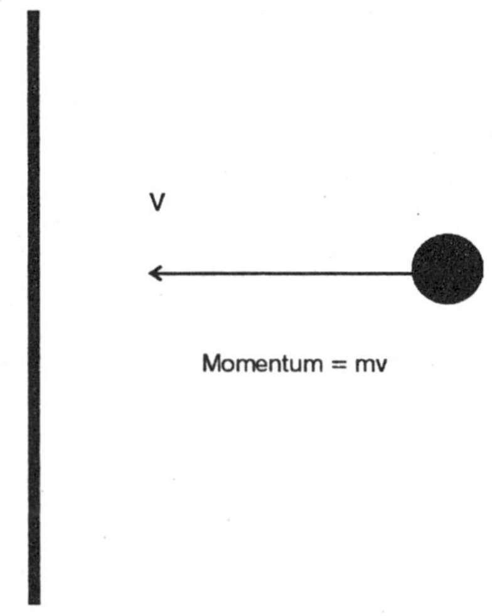

Fig. Collision of Gas Molecule with Container Wall

Since momentum is conserved in any collision, the change in momentum of the molecule is

momentum change = momentum after collision–momentum before= m(-v)–m(v) = -2mv

The change of momentum of the wall must therefore be 2mv to give a total change of zero. We now have the first term in equation:

momentum change/collision = 2mv

We take an intuitive approach to the second term. The number of collisions per unit time per unit area of wall should depend on 1) the number of molecules per unit volume in the container, N/V (the more there are, the more collisions there should be with the walls); and 2) the speed v at which a molecule moves (the faster the movement, the more collisions per unit time that should occur). Equation is based on these ideas:

collisions/time-area = (N/V)(v)

If we assume for simplicity that the container is cubical,

there are 6 walls over which these collisions must be spread. The collisions with a particular wall are then

$$\text{collisions/time-area} = Nv/6V$$

(Despite our restrictive assumptions about the shape of the container, equation turns out to be valid for a container of any shape!) We now multiply the expressions in equations to obtain the pressure:

$$P = (2mv)(Nv/6V) = Nmv^2/3V$$

Rearrangment gives equation.

$$PV = Nmv^2/3$$

This is as far as our molecular model takes us.

We are now at the interface between theory and experiment. Experiment (the ideal gas law) relates pressure and volume to temperature:

$$PV = nRT$$

Theory, equation, relates pressure and volume to the mass and speed of a gas molecule. To bring experiment and theory into correspondence, we must equate the right sides of the preceding two equations. In order for KMT to successfully explain the ideal gas law, it is necessary that

$$nRT = Nmv^2/3$$

Equation can be simplified by replacing n with N/N_o (N_o is Avogadro's Number); $mv^2/2$ by KE (the kinetic energy of an average gas molecule); and solving for KE. The result is

$$KE_{molecule} = 3RT/2N_o$$

We refine this equation by making two further realizations. First, since R and N_o are both constants of nature, so must their ratio be. It is symbolized k, and is called Boltzmann's constant. Its value is 1.381×10^{-23} J/K-particle. Second, we have recognized that the kinetic energy of a gas molecule changes frequently as it undergoes collisions. However, its average speed over time is constant. The refined equation is among the most important equations in science:

$$\text{Average } KE_{molecule} = 3kT/2$$

The average kinetic energy of a gas molecule, no matter what its chemical identity, depends only on the Kelvin temperature. This gives a deep insight into the concept of temperature: it is a measure of the average kinetic energies of the molecules of a substance, hence a measure of their average speed. As T is increased, molecules move faster; as it is lowered, they move slower. The temperature at which molecular motion ceases is absolute zero. Temperatures lower than absolute zero are impossible because a molecule may not have negative kinetic energy. Equation is simple and profound. Is there an experiment that can be done to test its validity? There is in fact a simple experiment that verifies equation in a rearranged form. If we replace the average kinetic energy of a molecule with the expression $m(v^2)_{avg}/2$, where $(v^2)_{avg}$ is the average of the square of the speed, and solve the resulting expression for $(v^2)_{avg}$, we obtain equation.

$$(v^2)_{avg} = 3kT/m$$

The square root of $(v^2)_{avg}$ is called the root mean square speed, and is for our purposes approximately equal to the average molecular speed. Making this equality gives equation.

$$v_{avg} = (3kT/m)^{1/2}$$

Multiplying both numerator and denominator of the argument on the right side of this equation by Avogadro's number N_o gives the useful variant in equation.

$$v_{avg} = (3RT/MM)^{1/2}$$

The implication of equations is that a heavy gas molecule moves more slowly than a light one, in a quantifiable way. The ratio of the speeds of the light (L) and heavy (H) molecules should be the square root of the inverse ratio of their molar masses:

$$(v_H/v_L)_{avg} = (MM_L/MM_H)^{1/2}$$

In 1846, Thomas Graham measured the rates of diffusion of various gases. Diffusion is the process by which gases disperse in space via random molecular motion. The results of Graham's experiments are summarized in equation rate diffusion of gas A/rate of diffusion of gas B =(density of gas B/density of gas A)$^{1/2}$

But we have previously seen that gas density is proportional to MM. If we make the logical assumption that a gas diffuses at a rate that is directly proportional to the average speed of its molecules, then Graham's Law of Diffusion is in exact agreement with equation.

Example. Calculate the ratio of the speeds of H_2 and CO_2 molecules at 25°C.

Solution. As long as temperature is the same for both molecules, its actual value is unimportant.

$v_{H2}/v_{CO2} = (MM_{CO2}/MM_{H2})^{1/2} = (44.0/2.02)^{1/2} = 4.67$

Example. Calculate the average speed of a molecule of nitrogen gas at room temperature, 25°C.

Solution. $v_{N2} = (3RT/MM)^{1/2} = (3(8.314\ J/K\text{-}mole)(298\ K)/(0.02801\ kg/mole))^{1/2} = 5.15\times10^2\ m/s$

This is 1150 miles per hour, or somewhere between the speeds of a commercial jet liner and a fighter jet.

Please note that in applying equation in Example, values of R and MM with appropriate units have been used, so that speed comes out with appropriate units. Use of R in units of, for example, atm-L, produces a speed with absurd units.

The curve is a plot of the fraction of molecules having speed s, f(s) (we use s for speed, v for velocity), versus the possible values of speed between zero and infinity. The curve has several noticeable features:

- The curve has the expected shape — low on both ends, indicating that few molecules have extremes of speed; and peaking in the middle, indicating that most molecules have speeds somewhere between the extremes.
- The curve is unsymmetrical because there is a definite lower limit (zero) but no definite upper limit to speed.
- The total area under the curve is the total number of gas molecules.
- The height of the curve at a particular speed (say s_1) is proportional to the fraction of molecules having speed s_1.
- The area under the curve between two speeds s_1 and s_2 (shaded in the figure) is the fraction of molecules haveing speeds in the range s_1 to s_2.
- The mathematical expression for f(s) obtained by Maxwell and Boltzmann is equation:

$$f(s) = (\text{constant})(s^2)\exp(-ms^2/2kT)$$

The s^2 term increases with increasing s and the

exponential term decreases with increasing s. Thus f(s) first increases, but then peaks and decreases with increasing s as the exponential term takes over. The speed at the maximum of the curve, s_{mp}, is the most probable speed, because the greatest fraction of molecules have it.

The Maxwell-Boltzmann Distribution Law. We have said that in a gas the molecules move randomly, and that the speed of a molecule changes frequently. In the mid 1800's, James Clerk Maxwell in England, and Ludwig Boltzmann in Austria, were concerned with using statistical methods to describe, precisely and mathematically, the distribution of molecular speeds. The result of their efforts is called the Maxwell- Boltzmann Distribution Law. They showed that at a given temperature, the molecular speed distribution follows a curve like that in Figure.

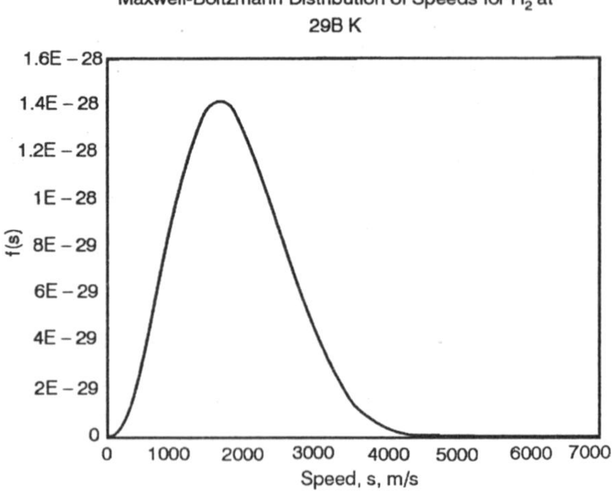

Fig. Maxwell-Boltzmann Distribution

- As T increases, the curve flattens and the maximum moves to higher speed, because the speeds of all molecules tend to increase with increasing T. The area under the curve stays constant, however, as long as the number of gas molecules is unchanged.

We now do two calculations using equation. First, we

calculate the most probable speed, s_{mp}. This is a good approximation to the average speed, but the average will be somewhat larger because the curve is biased to higher speeds. s_{mp} is the speed at which f(s) is maximum. At this maximum, the slope of the plot — the derivative of f(s) — is zero:

$$df(s)/ds = d/ds\,[(s^2)(\exp(-ms^2/2kT))] = 0$$

$$= ds^2/ds\,\exp(-ms^2/2kT) + s^2\,d[\exp(-ms^2/2kT)]/ds = 0$$

$$= 2s\,\exp(-ms^2/2kT) - s^2(ms/kT)\exp(-ms^2/2kT) = 0$$

Dividing by the exponential, we obtain

$$2s - ms^3/kT = 0$$

This is readily solved for s_{mp}:

$$s_{mp} = (2kT/m)^{1/2}$$

As expected, this is a bit smaller than the root-mean-square speed in equation.

Next we calculate the average KE of a gas molecule by a method of averaging; we multiply each possible kinetic energy by the number of molecules with that kinetic energy, and divide by the total number of molecules. The required mathematical procedure is integration.

$KE_{molecule}$ = Integral from 0 to Infinity of $f(s)(ms^2/2)ds$ $= 3kT/2$

Even if you are not yet able to carry out the integration, you should note that the result is the same as equation, which came from KMT. This is gratifying. It reinforces the validities of both KMT and the M-B Distribution Law.

Finally, we calculate the KE per mole of gas from the average $KE_{molecule}$:

$$KE_{mole} = N_o \times KE_{molecule} = 3RT/2$$

The kinetic energy of a mole of gas depends only on temperature. For this reason, kinetic energy of molecules is called thermal energy. Summary of the Interpretation of the Ideal Gas Law in Terms of Molecular Behaviour. The relationship between the macroscopic and microscopic views of gas behaviour can be summarized in several statements.

- Gases are compressible because they consist of small molecules that are far apart.
- Molecules of a gas are in constant motion, allowing gases to diffuse and to flow.

- Pressure results from collisions of gas molecules with the container walls.
- Pressure increases as volume decreases (Boyle's Law). A decrease in V increases the number of molecules per unit volume, leading to more molecular collisions with a unit area of the wall per unit time.
- An increase of temperature causes an increase in pressure (the Law of Gay-Lussac). KE of the molecules increases with T. Faster molecules cause an increase in pressure for two reasons:
  - There are more collisions with the wall per unit time;
  - Each collision exerts a greater force, since the molecule is moving faster.
- An increase in the amount (moles) of gas causes an increase in pressure at constant T (Avogadro's hypothesis). More gas means larger N/V. Higher N/V means more collisions per unit time and higher pressure.
- The ratio of diffusion rates of two gases is proportional to the square root of the inverse ratio of their molar masses (Graham's Law). This is a consequence of the average kinetic energy of a gas molecule being dependent only on T. At a particular T, a heavy molecule moves slower than a light one, but its larger mass offsets its smaller speed, giving the same kinetic energy. An important consequence of this is that a sample of gas with heavy molecules will exert the same pressure as a sample of gas with the same number of light molecules, as long as their temperatures and volumes are the same. Both samples satisfy the ideal gas law, independent of what their molecules are like. Although heavy molecules move slowly and strike the wall less often, they strike it with bigger mass, and cause the same pressure.

THE HISTORICAL ROLE OF GAS PHASE CHEMICAL REACTIONS

The first decade of the 19th century must have been both

exhilarating and frustrating for chemists. On the one hand, Dalton's atomic theory (1803) rationalized many known facts about chemical elements, compounds, and reactions. In particular, the theory provided an explanation for the laws of conservation of mass, definite proportions, and multiple proportions, and gave a rational basis for the idea of reproducible combining masses of the elements in compound formation (e.g., the combining masses of oxygen and hydrogen in forming water are in the ratio 7.94 to 1).

On the other hand, attempts to develop a scale of atomic masses were thwarted by the ignorance of chemical formulas. Chemists, led by Dalton, assumed that the formula for water was HO, and that water was formed from hydrogen and oxygen by equation:

$$H + O \rightarrow HO$$

This was the simplest assumption to make in the absence of definite knowledge of the formulas of elemental hydrogen, elemental oxygen, and water.

In 1809, Joseph Gay-Lussac conducted a series of experiments with gases that ultimately enabled the development of the atomic mass scale. Specifically, for reactions in which gases react to form other gases, he studied the relationships among the volumes of gaseous reactants consumed and the volumes of gaseous products formed. His result is of tremendous importance.

Gay-Lussac discovered that, as long as the experiments were conducted at a constant temperature and pressure, the volumes of reactants used and products produced always gave whole-number ratios. One of many reactions that he studied was that between hydrogen and oxygen to form water vapour. He found consistently and reproducibly that for every one volume of oxygen used, two volumes of hydrogen were required, and two volumes of water vapour were produced:

: 2 vol hydrogen + 1 vol oxygen → 2 vol water vapour

Gay-Lussac was not certain what to make of these results; however, Amadeo Avogadro was. He made the (to him) reasonable proposal that Gay-Lussac's results implied that equal volumes of gases, at the same temperature and pressure, contain equal numbers of molecules. In modern terms, we say that the volume occupied by a gas is proportional to the number of moles of gas present.

Operating from this assumption, he concluded and stated that Dalton's view of the hydrogen/oxygen reaction, expressed in 5-4-1, was incorrect, because it was not consistent with the two-to-one-to-two hydrogen/oxygen/water volume ratios that Gay-Lussac observed. He then went on to propose that hydrogen and oxygen occur as diatomic molecules, and that water in fact contains two atoms of hydrogen per atom of oxygen, so that the water formation reaction becomes:

$$2H_2(g) + O_2(g) \rightarrow 2H_2O(g)$$

This is the simplest proposal consistent with the combining volumes observed; as we know today, it is entirely correct.

GAS PRODUCTION VIA DOUBLE DISPLACEMENT REACTIONS

Three broad classifications of chemical reactions: electron transfer processes, proton transfer processes, and double displacement processes. We defined a double displacement reaction as one in which the positive and negative portions of the reactant molecules interchange to form products. It is appropriate on gases to focus briefly on double displacement reactions that produce a gas as one of the products. An important series of examples of such reactions involves metal carbonate compounds (called carbonate salts) and acids as reactants. The reaction of a metal carbonate with an acid is shown generically in equation.

$$MCO_3(\text{s or aq}) + HX(aq) \rightarrow MX_2(aq) + H_2CO_3(aq)$$

This reaction does not have a gas as one of its products. However, in a subsequent step, carbonic acid, H_2CO_3, decomposes to water and carbon dioxide, a gas that escapes in part from the aqueous solution in the form of effervescence:

$$H_2CO_3(aq) \rightarrow H_2O + CO_2(g)$$

A specific example of the acid-carbonate reaction. Here the net reaction, obtained as the sum of equations, is shown.

$$Na_2CO_3 + 2\,HCl \rightarrow 2NaCl + H_2O + CO_2$$

Acid-carbonate reactions are very important in at least two contexts. First, metal carbonates that have low solubility in water make up a substantial fraction of the earth's crust (limestone is an example). Deposits of carbonate rocks in contact with slightly acidic ground waters undergo reactions. In so doing, they serve to maintain the acidity level of the water at a fairly constant level, supporting aquatic plant and animal life.

Second, reactions are important in the context of acid rain, which results when nitric acid, HNO_3, and sulfuric acid, H_2SO_4, are produced in the atmosphere from SO_2, produced in coal combustion, and NO_2, a byproduct of the internal combustion engine. Limestone (primarily calcium carbonate) is widely used in construction, and has been used for centuries as a medium for sculptors.

Repeated and prolonged contact of building edifices and priceless works of art with acid rain results in their slow destruction as calcium carbonate reacts with nitric and sulfuric acids according to equations. Building edifices can be replaced; works of art cannot. Finally, dietary calcium tablets are made primarily of calcium carbonate. Contact of a stomach tablet with stomach acid (hydrochloric acid) causes the tablet to dissolve via reaction, after which the calcium can be utilized by the body.

$$CaCO_3(s) + HCl \rightarrow CaCl_2(aq) + H_2O + CO_2(g)$$

Metal sulfites (for example, Na_2SO_3) react with acids in analogous fashion to form aqueous sulfurous acid, H_2SO_3. This subsequently decomposes to water and sulfur dioxide:

$$Na_2SO_3(aq) + HNO_3(aq) \rightarrow NaNO_3(aq) + H_2SO_3(aq)$$

$$H_2SO_3(aq) \rightarrow H_2O + SO_2(g)$$

Example. A 500-mg calcium tablet dissolves in an excess of stomach acid. What volume of CO_2 is produced?

Solution. We begin with a balanced chemical equation for the reaction of stomach acid and the calcium tablet, which we assume to be pure calcium carbonate:

$$CaCO_3(s) + 2HCl(aq) \rightarrow CaCl_2(aq) + H_2O + CO_2(g)$$

To use the numerical information in the equation coefficients, we must work in moles:

moles $CaCO_3$ = 0.500 g/(100.086 g/mole) = 5.00×10^{-3} moles.

1 mole of CO_2 is produced for each mole of calcium carbonate that reacts. It follows that 5.00×10^{-3} moles of CO_2 are formed. The ideal gas law enables us to calculate the volume of CO_2. We assume the temperature in the stomach to be 98.6 °F, or 37 °C, and that the carbon dioxide is produced at 1 atm pressure. Then

$$V = nRT/P = (5.00\times10^{-3} \text{ moles})(0.08206 \text{ atm-L/mole-K})(310 \text{ K})/(1.00 \text{ atm})$$
$$= 0.127 \text{ L}$$

This volume of CO_2 is probably not sufficient to cause "indigestion."

Chapter 6

The Molecular Forces

INTRAMOLECULAR AND INTERMOLECULAR FORCES

The electrical force which is responsible for all attractive interactions between atoms, molecules, and ions. The electrical force is described quantitatively by Coulomb's Law. A system in which forces act can possess two types of energy. These are kinetic energy, KE, and potential energy, PE. The magnitude of the kinetic energy is a function only of temperature, as we have seen.

Potential energy arises from the action of the Coulomb force. We will discuss the various ways in which Coulomb forces operate in atomic/molecular systems, manifesting in both intramolecular and intermolecular contexts.

Intramolecular forces include the forces between atoms in covalent molecules; the forces between oppositely charged ions in ionic compounds; and the forces between identical atoms in metals.

Intermolecular forces include the forces that operate between one atom or molecule and another, or between the ion of one substance and the molecules of another.

Although there are differences among the various types of forces, they are all fundamentally similar: they give rise to potential energy between the interacting particles, described by the potential well.

We shall see that an interplay between the depth of this well and the thermal kinetic energies of particles is responsible for many of the physical properties of substances.

TYPES OF ENERGY AND THE COULOMB FORCE

An atom or a collection of atoms can possess energy of only two types. Kinetic energy (abbreviated KE) is energy of motion. A body of mass, m, moving with velocity, v, has kinetic energy given by equation.

$$KE = mv^2/2$$

When mass is expressed in kg and velocity in m/s, KE has units of kg m^2/s^2, or Joules (J). One Joule is defined as 1 kg m^2/s^2. The kinetic energy of atoms and molecules depends only on temperature, through the simple expression in 6-1-2.

$$\text{Average KE} = 3kT/2$$

Here Average KE is the average kinetic energy of a single atom or molecule; k is Boltzmann's constant; and T is the Kelvin temperature. Potential Energy (abbreviated PE) is energy of position. A rock poised on top of a hill, even though it is not moving, has the potential to move if given a small push. It therefore has potential energy. This potential energy is converted to kinetic energy as the rock rolls down the hill. All energy is either kinetic or potential. The total energy of any mechanical system is the sum of its kinetic and potential energies:

$$E = KE + PE$$

Potential energy is an abstract concept; it is difficult to understand how something can possess energy simply by virtue of its position. A weight rests on a platform at height h above the ground.

The weight is connected by a rope, across a pulley, to another platform that carries a load. If we remove the weight from the platform, it will fall, and in the process will lift the loaded platform. But in lifting the load a distance h, the weight has done work (a form of energy) on the load. To do work, the weight must use energy.

It possesses the necessary energy to do the work by virtue of its position at distance h above the earth. Thus the concept of potential energy presupposes a force of attraction or repulsion. The weight falls to earth because there is a gravitational force of attraction between the earth and the weight. To move the weight from the ground to the platform initially, we would have had to use an amount of energy sufficient to overcome this force of attraction.

The amount of energy used to lift the weight is stored in the earth-weight system as potential energy. We retrieve this energy by allowing the earth and weight to come together. Potential energy is a consequence of a force.

Our example has been based on the gravitational force because this is a familiar situation. There are 3 forces other than gravitation. One of these is the electromagnetic force. This is the force by which electrons are attracted and held by the nucleus in an atom.

The electromagnetic force is responsible for all interactions between atoms and molecules. It is far stronger than the gravitational force, which may be neglected in discussing atomic and molecular interactions (the electromagnetic force of repulsion between two electrons is 10^{42} times stronger than the gravitational force of attraction between them!).

The electromagnetic force operating at the atomic level is responsible for

- The attraction of electrons to the atomic nucleus;
- The bonding together of atoms in a molecule;
- The attraction between oppositely charged ions in an ionic compound;
- The attraction between molecules that leads to liquefaction and solidification;
- The flow of electrical current;
- Your inability to push your finger or hand through a wall.

The magnitude of the electrical force is given by Coulomb's Law, equation.

$$F = q_1q_2/WXr^2$$

q_1 and q_2 are the values of the two interacting electrical charges, in Coulombs, and r is the distance between them, in m. W is a constant that adjusts the units. Its value is 1.113×10^{-10} $C^2N^{-1}m^{-2}$. The force described by Coulomb's Law is responsible for all attractive forces between atoms, molecules, and ions. The potential energy associated with the Coulomb force is called the Coulomb potential energy. Because force is the negative derivative of potential energy with respect to distance (F = –dPE/dr), the Coulomb potential energy can be obtained by integration of Coulomb's Law.

$$PE = q_1q_2/WXr$$

When one of the charges is positive and the other negative, the potential energy is negative at all finite values of r.

Example. Calculate the Coulomb potential energy of an electron attracted by a proton at a distance of 52.9 pm. The charges of the proton and electron are 1.602×10^{-19} and -1.602×10^{-19} C, and $W = 1.112 \times 10^{-10}\ C^2J^{-1}\ m^{-1}$.

Solution.

$PE = q_1q_2/W \times r = -(1.602 \times 10^{-19})^2/(1.112 \times 10^{-10}\ C^2J^{-1}m^{-1})$ $(52.9 \times 10^{-12}\ m) = -4.36 \times 10^{-18}$ J

COULOMB FORCES AND THE POTENTIAL WELL.

Atoms, molecules, and ions exert attractive forces on one another because the atomic nuclei in one species attract the electrons in the other.

These are called molecular forces, and the potential energy that results from the action of these forces is called molecular potential energy.

We make several general statements about molecular forces and potential energy, and the relationship between them. The statements are equally valid for a pair of particles or a large collection of particles (here we are using the word, "particle", generically to represent an atom, a molecule, or an ion).

- Molecular forces are electrical;
- Molecular forces are attractive except at very close distances, and described by Coulomb's Law;
- The magnitude (strength) of the force between 2 particles varies with their distance of separation. The force is zero for an infinite separation, increases as the particles get closer, becomes a maximum when

the particles just touch, and becomes strongly repulsive when the centers of the particles are separated by less than one particle diameter, due to interpenetration and repulsion of electron clouds. The force has two components: one operates only at very small distance and is repulsive; the other acts at all distances and is attractive. Thus

$$F = q_1q_2/WXr^2 + B$$

B is a positive, repulsive force that operates only at very small distance.

- Action of the molecular forces draws the particles together (tends to cause them to aggregate) and results in a decrease in the potential energy of the pair (or collection) of particles. The force becomes stronger and the potential energy less until the particles touch.

  At distances less than 1 particle diameter, the force is repulsive and PE increases again. For all types of molecular forces, potential energy varies with distance of molecular separation, r. The plot is shaped like a valley, or well. It is usually called a *potential energy well.* The depth of the well is the energy that must be added to increase the distance of separation between particles from r_1 to r_3 or greater.
- There are several types of molecular forces, which we will discuss shortly. All of them are characterized by a potential well of similar shape. However, the depth of the well increases with the strength of the forces.

Region. The particles, shown as two small circles just above the horizontal axis, are separated by large distances. Taking 2×10^{-10} m as a typical particle radius, the distances of separation in region 1 are 10 or more times larger. At large separation distance, molecular forces are nearly zero. For convenience, we take the potential energy to be zero when the forces are zero, so the PE curve coincides with the horizontal axis at large r.

This situation of very large separation between atoms,

molecules, or ions is a state of *deaggregation*. The gas phase of a pure substance is an example of a deaggregated state.

Region. The particles have approached closely enough to exert noticeable forces on one another. Action of the forces lowers the potential enrgy. The closer the particles approach (i.e., the more aggregated they become), the stronger the forces and the lower the PE. The particles just touch. At this point, attractive forces are maximized and potential energy is at a minimum. This is an aggregated state. The solid phase of a pure substance is an example of an aggregated state.

Region. PE increases very rapidly as r decreases. In this region, the somewhat deformable electron clouds of the particles are being pushed into one another.

Since electrons repel one another, two particles that have interpenetrated exert a repulsive force on one another, and tend to fly apart. Region shows that when particles are forced together so that their centers are separated by less than one molecular diameter, repulsive forces, hence PE, rise very rapidly. This is the force that keeps you from falling through the floor.

The attractive forces are maximized and the PE minimized when the atoms, molecules, or ions just touch, at the bottom of the well. The distance corresponding to the potential energy minimum is the distance of closest approach. We have called this the aggregated state. In an ionic compound, such as KBr, this corresponds to a situation in which positive ions are surrounded by negative ions, and vice versa.

In a covalent compound, such as H_2O, the bottom of the well represents the situation in which the hydrogen atoms are covalently bonded to the oxygen atom, the bonds are separated by an angle of 104°, and the distance between the H and O atoms corresponds to r at the bottom of the well. A collection of He atoms or methane molecules can maximize the contact between themselves by aggregating such that a particular molecule touches as many other molecules as possible.

This organization is referred to as close packing, and

characterizes the solid state of a pure substance. PE is minimized in the aggregated state.

Thus shows how potential energy changes as a substance is converted from the deaggregated to the aggregated state or vice versa. Notice that potential energy increases when an aggregated state is converted to a deaggregated one:

$$\text{Aggregated} \rightarrow \text{Deaggregated}$$
$$\text{DPE} > 0$$

This is an idea that has broad applicability; we will encounter it many times.

The phases (solid, liquid, and gas) of a pure substance fit the aggregation/deaggregation model very well. In the solid phase, the fundamental units of a substance are packed together in an orderly arrangement that minimizes potential energy. This is the aggregated state, in which intermolecular attractive forces are optimized. In the gas phase, the fundamental units are widely separated and experience only very weak forces from one another.

This is the deaggregated state. We might ask where the liquid state is on the diagram. From macroscopic observation, we know that the molar volume of a substance, e.g. water, in the liquid phase is very little different from the molar volume of its solid phase. These two aggregated (or condensed) phases have molar volume roughly 1000 times smaller than that of the gas phase. The liquid phase has distance of molecular separation and potential energy only somewhat greater than those of the solid, at the point labelled r_2.

The condensed (aggregated) phases of matter and the conversions between the gas, liquid, and solid phases. Prior to that, it is important that you have an understanding of the types of forces that operate within and between molecules, and their relative strengths. This will enable you to understand, for example, why some substances exist as high-melting solids under normal conditions, while others exist as low melting solids, liquids, or gases.

TYPES OF MOLECULAR FORCES

The nature and structure of a substance determine the type

and strength of molecular forces that operate within it. We divide such forces into two broad categories. Intramolecular forces operate within the molecules or fundamental units of a substance. These are the covalent bonds within a molecule of a molecular substance, the forces between ions in an ionic compound, and the forces between atoms in a metal. Intermolecular forces operate between, rather than within, the molecules of a covalent substance; the atoms of a monatomic element; or the ions of one substance and the molecules of another. As a rule, intramolecular forces are much stronger than intermolecular forces. Both types of force, however, give rise to potential energy and a potential well. A listing of the important forces, arranged within these categories, follows:

Intramolecular Forces

- Attraction between positive and negative ions in a crystal of an ionic compound.
- Covalent bonds linking atoms in a network structure, such as that of silicon dioxide (sand).
- Covalent bonds in molecular substances.
- Metallic bonds

Intermolecular Forces

- Attractions exerted by one molecule of a molecular substance on another, such as the force of attraction between water molecules in ice.
- Attractions between atoms of the noble gas elements, helium through radon.
- Attractions between molecules of one substance and molecules of another, as when two liquids are mixed, or a molecular solid such as sugar is dissolved in a liquid.
- Attraction between molecules of one substance and ions of another, as when an ionic compound dissolves in a liquid.

In succeeding sections, we will discuss each of these molecular forces, and their relative magnitudes. We will begin with intramolecular forces.

INTRAMOLECULAR FORCES

THE IONIC BOND

When two elements of widely different electronegativities combine, the result is an ionic compound. An example of an ionic compound is potassium bromide, KBr, in which the electronegativity difference between bromine, in group 17, and potassium, in group 1, is 2.14. Ionic bond formation involves transfer of an electron from the potassium atom, with its single valence electron and core charge, Z_{core}, of 1 unit, to the bromine atom, with its seven valence electrons and large Z_{core} of 7 units. The electron transfer generates a full valence shell for each species.

The ionic bond can be understood in terms of three energy terms. To remove an electron from the potassium atom to generate the potassium ion requires input of the first ionization energy of potassium. The process is shown below:

$$K \rightarrow K^+ + e^- \quad I_1 = 420 \text{ kJ/mole}$$

I_1 for potassium is relatively small due to the minimum core charge found for group 1 elements. To add an electron to the bromine atom to form the bromide ion releases an amount of energy equal to the first electron affinity of bromine.

$$Br + e^- \rightarrow Br^- \quad -A_1 = -325 \text{ kJ/mole}$$

The first electron affinity for bromine is substantial, because the large core charge of the bromine atom exerts a reasonable force on the extra electron. Despite this, to accomplish the electron transfer from potassium to bromine requires that 420–325 = 95 kJ/mole of energy be supplied. In other words, the electron transfer is substantially unfavorable! Thus we find that, contrary to a commonly held view of ionic bond formation, potassium does NOT "like" to transfer an electron to bromine; the process is unfavorable, and is not responsible for the strength of the ionic bond. What, then, is?

Generation of positive and negative ions results in a strong attraction between them, according to Coulomb's Law. A single positive and negative ion are shown coming together to form an ion pair in reaction below.

$$K^+ + Br^- \rightarrow K^+Br^- \text{ (ion pair)}$$

Formation of one mole of ion pairs releases 589 kJ of energy, much more than was required to generate the ions. We conclude that it is the Coulomb attraction between oppositely charged ions that drives the formation of the ionic bond; this happens despite the need to form ions, which is unfavorable. In fact, ionic compounds do not consist of ion pairs of the type shown in reaction above. Rather, they consist of extended, highly organized aggregates in which positive and negative ions alternate. Figureshows both the ion pair and the lattice of potassium bromide.

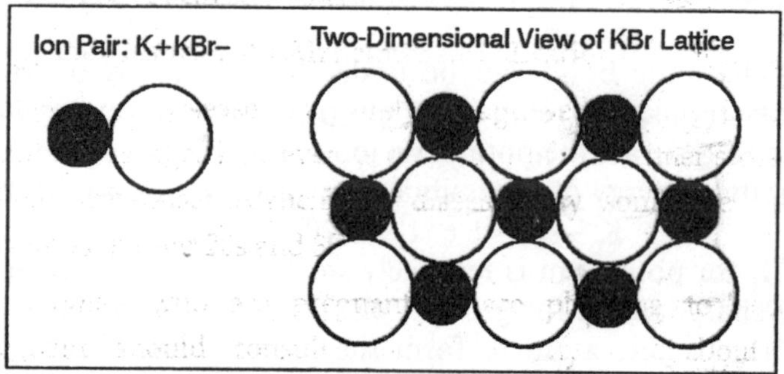

Fig. Potassium Bromide, Ion Pair and Lattice

The lattice is more stable than a collection of ion pairs because it surrounds each ion with ions of the opposite charge. This maximizes the attractive forces between positive and negative ions. In the ion pair, the positive charge of the potassium ion is offset on one side by the bromide ion, but is still exposed on all other sides.

This exposed positive charge attracts bromide ions of other ion pairs, causing the ion pairs to condense to a lattice. Formation of a lattice containing 1 mole each of potassium and bromide ions as in reaction releases 672 kJ of energy. This number, about 50 per cent larger than the amount released in formation of the equivalent amount of ion pairs, represents the depth of the potential well for potassium bromide.

$$K^+(g) + Br^-(g) \rightarrow KBr \text{ (lattice)}$$

The energy change that occurs when 1 mole of ionic substance in the solid lattice is formed from separated positive and negative ions in the gas phase is called the lattice energy, U, of the ionic compound. The lattice energy is given by an expression having the form of:

$$U = KZ^{+}Z^{-}/r \text{ m}$$

Here K is a collection of physical constants, Z^{+} is the cation charge, Z^{-} is the anion charge, and r is the distance of separation of the cation and anion in the lattice. Because the charge of the anion, Z^{-}, is negative, the lattice energy is a negative quantity, as expected. This means simply that the solid lattice is more stable than the separated gas phase ions. Note that this equation is essentially the expression for Coulomb potential energy. Without doing any calculations at all, we can draw some important conclusions based on reaction.

- The potential well for an ionic compound of a 2+ cation with a 2- anion will be about 4 times deeper than that for a 1+/1- ionic compound, like KBr, due to the product of charges in the numerator of equation. Knowing the lattice energy for KBr allows us to estimate the lattice energy for MgO: about -2600 kJ/mole
- The potential well for an ionic compound will be deeper when cation and anion are small than when they are large. Small cation and anion means small distance of separation, r, leading to a more negative U. The lattice energy for LiF is substantially more negative than that for KBr, despite the fact that both are 1+/1- compounds, because Li+ and F- are smaller ions.

Lattice energies for a number of common ionic compounds are given in Table.

Table. Lattice Energies of Ionic Compounds

Charge Type	*Substance*	*U, kJ/mole*
1 + /1–	LiF	–1009
	NaCl	–774
	KBr	–672
	CsI	–584
2 +/2–	MgO	–2585
	BaO	–1978

Example. The lattice energy for NaCl is -774 kJ/mole. The radii of the sodium and chloride ions are 116 pm and 167 pm respectively. Estimate the lattice energy for CsI, which has ion radii of 181 and 206 pm.

Solution. NaCl and CsI are both 1+/1- ionic compounds. They should differ in lattice energy only through the r term of 6-4-5. Because the distance of separation is larger in CsI, the lattice energy should be less negative. We assume that r can be approximated by adding the radii of the ions. Thus for NaCl, r = 283 pm; for CsI, r = 387 pm. The lattice energies should be inversely related to the distance of separation of the ions:

$$U(CsI)/U(NaCl) = r(NaCl)/r(CsI)$$
$$U(CsI)/(-774) = 283/387$$
$$U(CsI) = -570 \text{ kJ/mole}$$

This compares reasonably well with the value in Table.

Example. Arrange the following ionic compounds in order of increasing intramolecular forces: KI, LiCl, MgO, CaO, Al_2O_3.

Solution. KI and LiCl are 1+/1- ionic compounds. MgO and CaO are 2+/2- ionic compounds. Finally, Al_2O_3 is a 3+/ 2- compound. Based on ion charges and equation, we can arrange these in three categories:

KI, LiCl
MgO, CaO
Al_2O_3

The 1+/1- compounds have the least negative lattice energies. That for Al_2O_3 should be most negative, because the ion charges are largest. Within each category, the compound with the largest ions has the least negative lattice energy. This reasoning leads to the following order:

$$U(KI) > U(LiCl) > U(CaO) > U(MgO) > U(Al_2O_3)$$

Here the greater than sign (>) is applied in the algebraic sense, in which numbers are considered to become smaller as they get more negative. In this convention, -50 is greater than – 100.

There are very few compounds that are legitimately considered to be ionic; that is, in which electron transfer from metal to non-metal is complete. They are limited to compounds of the group 1 metals (Li, Na, K, Rb, Cs) with fluorine, chlorine, and oxygen. Compounds of the group 2 metals with oxygen, sulfur, and the halogens form typical ionic lattices, and are usually considered to be ionic compounds. However, there is considerable covalent character in the bonding.

Based on our discussion of the ionic bond in this section, it is understandable why ionic compounds are invariably solids at room temperature, with high melting and boiling points. The strong, non-directional Coulomb attractions acting between oppositely charged ions in the lattice can be overcome only at very high temperature.

The Covalent Bond

Two atoms of relatively large and similar electronegativities (i.e., non-metals) tend to form covalent compounds rather than ionic compounds. In covalent compounds, atoms share rather than transfer electrons to achieve full valence shells, because both atoms have similar substantial values of the core charge. Large Z_{core} tends to make electron transfer unfavorable. The shared electrons are paired, and are usually considered to be localized between the bonded atoms. In this section, we will discuss the chemist's view of the covalent bond. Two separated chlorine atoms, each with 7 valence electrons distributed in the 3s and 3p valence orbitals.

Each Cl atom has a single odd electron in a 3p orbital. It is these electrons that will be paired and shared in the covalent bond. For the two atoms to bond by sharing electrons, it is necessary that they establish a mutual region in which the shared pair resides. They accomplish this by merging, or overlapping, the atomic orbitals containing the electrons to be paired. Because these atomic orbitals have distinct spatial directions, the bonded atoms must maintain a definite orientation relative to one another in order to preserve the overlap of orbitals. Thus covalent bonds are directional.

The operation of Coulomb's Law in the covalent bond is perhaps less obvious than in the ionic bond. The two chlorine nuclei and the two electrons of the covalent bond, and indicates the various Coulomb forces that operate in this collection of centers. There are several interactions: attraction of nucleus 1 for electron 1; attraction of nucleus 2 for electron 1; attraction of nucleus 1 for electron 2; attraction of nucleus 2 for electron 2; repulsion of nuclei 1 and 2; and repulsion of electrons 1 and 2. Note, however, that because the bonding electrons are required to be between the nuclei so that they can occupy the overlapping atomic orbitals, the nucleus-electron attractive forces are maximized, and the nuclear-nuclear repulsions are minimized, because the electrons shield the two nuclei from each other. Thus the strength of the covalent bond is attributable to the simultaneous attraction of both electrons to both nuclei.

The lengths and strengths of covalent bonds—the position and depth of the covalent bond potential well—vary with the bonded atoms. In general, the strengths are substantial and comparable with those of ionic bonds. Thus to break all of the covalent bonds in a molecule of methane, CH_4, requires the input of 1662 kJ of energy. Methane lies in a deep potential well relative to the separated atoms. Generally speaking, the following generalizations hold for covalent bond strengths.

- Small atoms form relatively short bonds, and large atoms form relatively long bonds.
- Short bonds tend to be stronger than bonds formed by larger atoms because the bonding electrons are closer to the bonded nuclei. For example

Bond	*Length*	*Strength*
Cl–Cl	198 pm	240
Br–Br	228 pm	190
I–I	266 pm	149

- The word "tend" is emphasized here, because there are exceptions. Notably, single bonds of the small atoms, N, O, and F are very weak.
- Multiple bonds are stronger than single bonds. For example, the single, double, and triple bonds between carbon and oxygen have the following strengths:

C – O 358 kJ/mole

C = O 799 kJ/mole

C(3)O 1071 kJ/mole

This seems reasonable, because there are more electrons localized between the nuclei and attracted by both. Strengths and lengths of some typical covalent bonds are presented in Table. A more complete set of covalent bond strengths is found in Table below.

Table. Strengths and Lengths of Covalent Bonds

Bond	*Bond Energy (Strength), kJ/mole*	*Bond Length, pm*
Single Bonds		
H – H	432	74.2
H – F	565	91.8
H – Cl	428	127.4
H – C	411	109
H – N	386	101
H – O	459	96
C – C	346	154
C – F	485	135
C – Cl	327	177
C – O	358	143
C – N	305	147
N – N	167	145
N – O	201	140
N – F	283	136
O – O	142	148
O – F	190	142
Multiple Bonds		
C = C	602	134
C = O	799	120
C = N	615	
N = N	418	125
N = O	607	121
O = O	494	120.7
S = S	425	188.7
C(3)C	835	120
C(3)O	1072	113
C(3)N	887	116
N(3)N	942	109.8
P(3)P	481	189
As(3)As	380	

Example. Arrange the following single bonds in order of increasing length. Then arrange them in order of increasing strength.

H – O, O – Cl, Cl – Cl, H – Cl

Solution. Longer bonds are formed by larger atoms; atoms become larger as we descend the periodic table. The sizes of the atoms involved here are ordered as follows:

Cl > O > H

The bond lengths are then expected to be in this order: H-O < H-Cl < O-Cl < Cl-Cl. The strengths of bonds tend to decrease with increasing length. Thus: Cl-Cl < O-Cl < H-Cl < H-O.

Although simplified, this section gives us an appreciation for the origin and strengths of covalent bonds. We now turn to a problem with our view of the covalent bond having to do with molecular stereochemistry.

The prediction and rationalization of molecular stereochemistry (shape): the VSEPR theory. Thus VSEPR predicts the tetrahedral geometry for methane, in exact agreement with experiment.

The interpretation of the covalent bond in terms of overlap of atomic orbitals of the bonded atoms requires that we understand molecular shape in terms of the distribution in space of the atomic orbitals of the central atom, which are presumably overlapped with suitable terminal atom orbitals in covalent bonds.

We pose this question: are the observed shapes of molecules consistent with the distribution in space of the atomic orbitals of the bonded atoms? If we ask this question of methane, the answer must be NO.

On one hand, the C-H bonds are known to make 109.5 ^{o} angles with one another. On the other hand, the s and p valence orbitals of carbon have spatial orientations that are not consistent with this angle.

The s orbital is spherical, hence non-directional; and the p orbitals make mutual angles of 90 ^{o}. In fact, the bond angles in most molecules in which 3, 4, or 5 electron groups surround the central atom are inconsistent with this 90 ^{o} orbital orientation. This presents a problem for our view of the covalent bond.

This problem was recognized in the 1920s, when the modern quantum theory was applied to the hydrogen atom and the spatial shapes of the atomic orbitals were discovered. Tetrahedral stereochemistry at carbon in methane had been proposed in 1874 by.

Vant Hoff on the basis of the number of isomers produced when hydrogen atoms were replaced by different atoms. At that time, though, people knew nothing of atomic orbitals.

Thus the new quantum theory was faced almost immediately with a challenge: explain the shape of methane. This problem was tackled and resolved by Linus Pauling, who made during his life almost innumerable substantial contributions to science.

Pauling proposed that the carbon atom uses hybrid orbitals to form four identical bonds to hydrogen atoms, directed to the vertices of a tetrahedron. His theory of hybrid atomic orbitals is still widely used today to rationalize (not predict) molecular stereochemistry. Applied to methane, the theory goes as follows.

In its most stable electron configuration, the carbon atom. The s orbital is fully occupied with an electron pair, and two of the three p orbitals are singly occupied. Based on this arrangement, we might expect carbon to form two bonds using the two odd electrons. However, this would fail to give it the octet.

Thus we imagine that carbon moves, or "promotes", one of the 2s electrons to the empty p orbital. This prepares it to form 4 bonds.

However, bond formation obviously does not occur using the pure atomic orbitals, because these can not produce a tetrahedral stereochemistry when overlapped with 1s orbitals of hydrogen atoms. Pauling proposed that the carbon atom creates from its 4 valence orbitals four new, equivalent orbitals called "hybrids."

The word hybrid means the same thing here as in

agriculture: a blend or combination of two strains of a species. He called these sp^3 hybrids to indicate that they are composed of a mixture of a single s and three p orbitals. Hybrid formation is shown schematically in Figure.

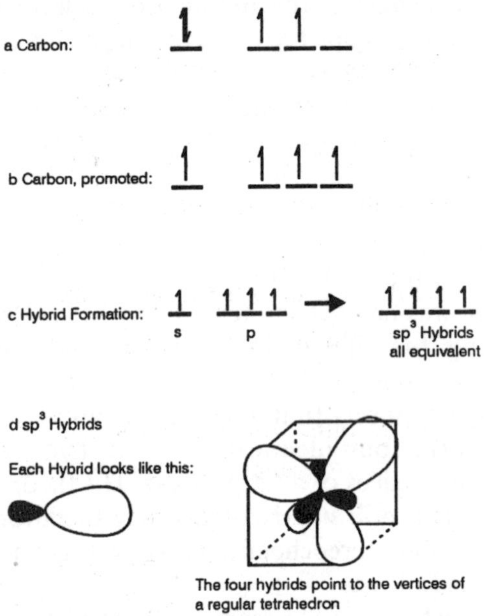

Fig. The Electron Configuration of Carbon

Indeed, he showed mathematically that it is possible to combine the four s and p atomic orbitals to produce four new orbitals, and that these new orbitals are directed to the corners of a tetrahedron. The four hybrids are shown in Figure above. Pauling thus "squared" fact with quantum theory by proposing hybrid orbitals. Let's apply these ideas to a few more situations.

The phosgene ($COCl_2$) molecule is pictured in Figure. It has been experimentally determined that the molecule is trigonal planar (as expected from VSEPR), with a double bond between carbon and oxygen. How is the trigonal planar stereochemistry, involving 120 ° bond angles, to be rationalized with the 90 ° spatial distribution of carbon 2p orbitals? Pauling showed that the 120 ° bond angle results if one assumes that C hybridizes the s and two of the three p orbitals to produce sp^2 hybrids. Mathematically, the mixing of an s and two p's does indeed produce three equivalent new orbitals, directed to the corners of an equilateral triangle.

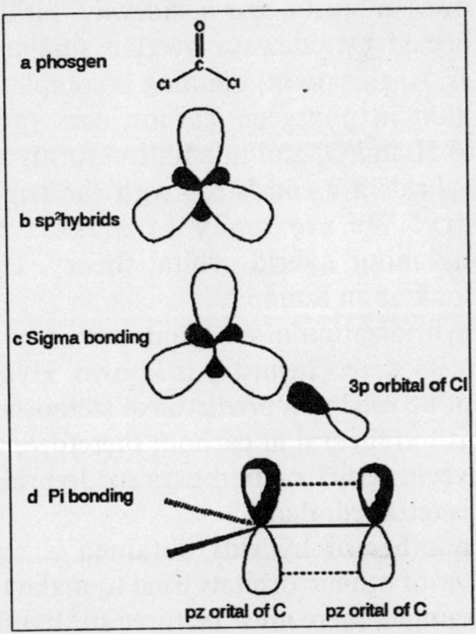

Fig. Bonding in Phosgene

The plane in which the 3 hybrids lie depends upon which two p orbitals are used in the process. By convention, chemists normally choose to use px and py in the hybrids, which requires that the hybrids lie in the x-y plane of a Cartesian coordinate system. The pz orbital, not used in the hybrids, retains its usual shape and orientation along the z axis. Note, however, that it still contains an odd electron.

Once the hybrids are formed, each containing a single electron, carbon forms covalent bonds to the two hydrogen atoms and the oxygen atom. Each bond is formed by overlap of one carbon hybrid with one atomic orbital of the bonded atom (H or O). In forming these three bonds, orbitals are overlapped along the line joining the bonded nuclei. Bonds involving overlap of this type are called *sigma bonds*. At this point, bonding is not complete because carbon shares in only

7 electrons, as does oxygen. Each can complete the octet by forming one more bond, using the p orbitals oriented along the z axis.

Because these do not point at one another, but instead are parallel, overlap must occur in sideways fashion.

Bonds formed by sideways overlap similar to this are called *pi bonds*. At this point, bonding is complete.

We say that in phosgene, carbon uses sp^2 hybrids to sigma bond to H and O, and in addition forms a pi bond to O. The sp^2 hybrids are consistent with the trigonal planar sterochemistry. We are ready to make a few useful generalizations about hybrid orbital theory. Having done that, we will look at an example.

- The hybridization of an atom cannot be determined unless its stereochemistry is known. Hybrid orbitals cannot be used as a predictor of stereochemistry. CH_4 is tetrahedral at carbon; therefore, carbon uses sp^3 hybrids.NOT carbon uses sp^3 hybrids; therefore CH_4 is tetrahedral at C.
- The number of hybrids obtained is equal to the number of atomic orbitals used to make them. Thus, for example, there must be three sp^2 hybrids because three atomic orbitals (one s and two p) are used.
- An atom uses hybrid orbitals for its sigma bonds and lone pairs. Thus the number of hybrids required is equal to the sum of the number of lone pairs and sigma bonds:
- 4 sigma bonds + 0 lone pairs requires 4 hybrids, thus sp^3 $(1 + 3 = 4)$
- 3 sigma bonds + 1 lone pair requires 4 hybrids, thus sp^3 $(1 + 3 = 4)$
- 2 sigma bonds + 2 lone pairs requires 4 hybrids, thus sp^3 $(1 + 3 = 4)$
- 3 sigma bonds + 0 lone pairs requires 3 hybrids, thus sp^2 $(1 + 2 = 3)$
- 2 sigma bonds + 1 lone pair requires 3 hybrids, thus sp^2 $(1 + 2 = 3)$
- 2 sigma bonds + 0 lone pairs requires 2 hybrids, thus

sp (1 + 1 = 2)

- Unhybridized p orbitals are used for pi bonds. The number of pi bonds equals the number of unhybridized p orbitals.

The flow of logic in determining hybrids used by the central atom is as follows. First, count the number of electron groups around the central atom; then apply VSEPR to determine the stereochemistry at that atom.

The number of hybrids is the same as the number of electron groups counted for VSEPR. A number of atomic orbitals equal to the required number of hybrids is used in forming the hybrids. The sum of the exponents in the hybrid designation indicates the number of hybrids of that type, and the number of atomic orbitals of each type used to make them.

Example. For each molecule, determine the Lewis structure; use VSEPR to determine molecular shape; and state the hybrid orbitals used by the central atom in bonding to the terminal atoms: $SiCl_4$; NH_3; BF_3; CO_2.

Solution. Lewis structures, then

Molecule	*#groups (hybrids)*	*Stereochemistry*	*Hybrids*
$SiCl_4$	4	Tetrahedral	sp^3
NH_3	4	Trigonal pyramidal	sp^3
BF_3	3	Trigonal planar	sp^2
CO_2	2	Linear	sp

Note that the number of pi bonds in BF_3 and CO_2 is the same as the number of unhybridized p orbitals on the central atom.

It is relatively simple to extend the hybrid orbital concept to rationalize the stereochemistries of molecules in which the central atom has an expanded valence shell (more than 8 electrons).

Essentially, this is done by incorporating in the hybrids as many valence d orbitals as required to give the necessary number of hybrid orbitals.

Example. Determine the hybrid orbitals used by the central atom in each molecule: PF_5, ClF_3, $XeOF_4$.

Solution.

Molecule	*# Electron Groups*	*Shape*
PF_5	5	Trigonal bipyramidal
ClF_3	5	T
$XeOF_4$	6	Square Pyramidal

Based on the numbers of groups, we deduce the hybrids. Five groups require 5 hybrids, which in turn require 5 atomic orbitals: one s, three p, and one d. Thus sp^3d. Six groups require 6 hybrids, which require the input of 6 atomic orbitals: one s, three p, and two d. Thus sp^3d^2. It can be mathematically demonstrated that 5 sp^3d hybrids point to the vertices of a trigonal bipyramid; and that 6 sp^3d^2 hybrids point to the vertices of an octahedron.

Before we leave the covalent bond and the hybrid orbital idea, something must be made clear. What is experimentally known is that carbon and hydrogen react to form methane, and that the methane molecule involves four carbon-hydrogen bonds that are perfectly tetrahedrally distributed in space. We introduced the ideas of electron promotion and orbital hybridization above as a way of rationalizing for ourselves, in terms of the output of quantum theory, what is experimentally observed.

Thus promotion and hybridization are mental constructs. It is important that you realise that the carbon atom does not actually do these things. It is simply convenient for us to think of the carbon atom as doing these things.

There are numerous elements and compounds that consist of huge numbers of atoms covalently bonded together to give a large, 2- or 3-dimensional network. These substances are therefore similar to ionic compounds in that they do not consist of small, discrete molecules; however, they do not consist of discrete ions either.

Examples of such substances are silicon dioxide (SiO_2), arsenic trioxide (As_2O_3), carbon in the diamond form, and phosphorus in the red form.

The large numbers of covalent bonds in these structures leads to potential well depths comparable to those found in ionic compounds. Most such materials exist as solids at ordinary temperature, and melt and boil only at very elevated temperature.

Ionic and Covalent Bonding—A Continuum

The bond between two identical non-metal atoms, as in Cl_2, is purely covalent; that is, each atom has an exactly equal share in the bonding electron pair. Pure covalent bonds exist in the elemental forms of many of the non-metals (for example, all of the diatomic halogens of group 7; O_2 and S_8 of group 6; N_2 and P_4 of group 5; C and Si of group 4; and others). Bonds between different non-metal atoms are covalent, in that they involve electron sharing, but polar, meaning that they have some ionic character. The bonds between chlorine and the other elements of period 3 are shown below. The transition from pure covalent to polar covalent to ionic as we move from right to left in the period is clear.

Na-Cl — Mg-Cl — Al-Cl — Si-Cl — P-Cl — S-Cl — Cl-Cl
ionic — ionic — ionic — polar cov — polar cov — polar cov — cov

Although the Na-Cl and Mg-Cl bonds are usually considered to be ionic, they are not purely ionic; electron transfer from the sodium or magnesium atom to the chlorine atom is not complete. There is no such thing as a pure ionic bond. The process of electron transfer from a metal atom to a non-metal atom generates a positive ion and a negative ion. The positive ion is electron deficient, and attracts an electron pair of the negative ion strongly. Thus the very act of electron transfer creates a disparity that must then be offset by partial electron sharing. If the anion is one of the very electronegative non-metals (fluorine or oxygen), it can resist the pull of the cation effectively, and hold on to the electron pair.

Anions other than F^- and O^{2-}, however, cannot completely resist the attraction of the cation for their lone pairs. These atoms share a lone pair with the cation to form a bond that, although it is very polar, is nonetheless partially covalent. As a general rule, the higher the charges on the cation and anion, the more covalency in the bond. A highly charged cation (3^+ or 4^+) is small and compact, and exerts a powerful attractive force on electrons. A highly charged anion (2^-, 3^-) holds its extra electrons weakly and readily shares them to some extent with the cation. The following series of compounds illustrates nicely the evolution from largely ionic to covalent bonding as cation charge increases:

Compound	*Physical State, 298 K*	*Melting Point °C*	
NaCl	Solid	801	
$MgCl_2$	Solid	714	
$AlCl_3$	Solid	190	
$TiCl_4$	Iiquid	-25	

The decrease in melting point indicates increasing covalency in the bonding. Titanium tetrachloride, $TiCl_4$, is a liquid at room temperature, indicating that it is molecular, not ionic.

The Metallic Bond. Elements of similar but relatively low electronegativities (i.e., metals) form metallic bonds. It is these bonds that are operative in a typical metal and that are responsible for metallic properties: reflectivity, conductivity, malleability, and strength. Metallic bonds form between atoms that have fewer valence electrons than they have valence orbitals. We shall try to develop a picture of the metallic bond by considering sodium metal, which exists as a very low melting, highly reactive solid at room temperature. Figure shows the Lewis dot symbols for two sodium atoms.

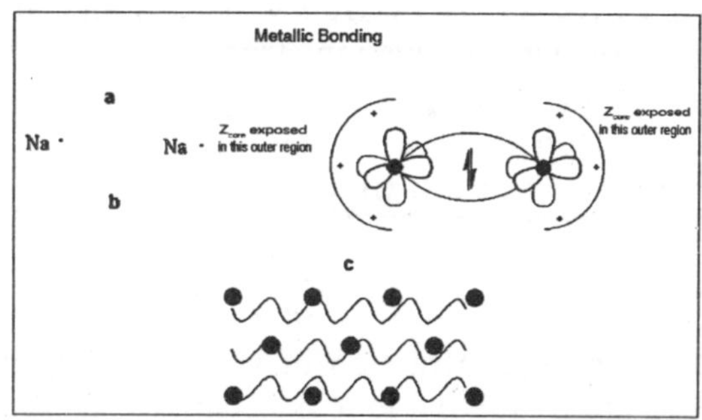

Fig. Metallic Bonding

The single odd electron is in the 3s valence atomic orbital

of the sodium atom. At first glance, it might be expected that two sodium atoms, each with a single unpaired valence electron, would form a single covalent bond by overlapping the 3s valence orbitals and pairing the odd electrons in the resulting common region of space. There are problems with this bonding situation however.

First, bonding of this type would give each sodium atom only a duet of electrons, nowhere near the required octet. Each atom would have three empty 3p orbitals protruding from it into space. Second, pairing the 3s valence electrons localizes them in the region between the two nuclei.

They no longer have probability of being found in the peripheral regions opposite the bond. Because these are the only valence electrons, the regions marked with + symbols in the figure are exposed to the full core charge (+1) of the sodium atom. (The + signs in the figure do not represent actual positive charges; instead, they are intended to indicate that another atom located in this region would experience the core charge of the sodium atom near it).

Here is the key point: another sodium atom in the region of the bonded pair would be drawn, via its valence electron, to the exposed core charge of the 2-atom cluster. It would contribute its valence electron to a bond with one or the other of the pair. The sodium atoms are therefore driven to aggregate, using their relatively few valence electrons to bond together as many nuclei as possible via the Coulomb attraction. The aggregation is very regular in nature, resulting in an ordered arrangement of metal atoms called a *lattice*. As a result of aggregation, each sodium atom uses all of its valence orbitals in forming partial bonds with a number of neighbors.

Contrast this with the covalent bonding between two chlorine atoms. The covalent bond is again localized, but the remaining valence electrons are distributed evenly around each chlorine atom, effectively shielding the core charge from chlorine atoms in other molecules. Further, all of the s and p valence orbitals are full. In order to aggregate as sodium does, chlorine would have to use either the 3d orbitals or the orbitals from the n = 4 main shell. But the core charge experienced by an electron in either of these orbital sets is essentially zero (17-17 = 0).

Consequently, chlorine is limited in its aggregation to the formation of covalently bonded diatomic molecules. As we will discuss shortly, there is only a very weak tendency for the diatomic molecules to aggregate via intermolecular forces.

Interestingly, although hydrogen, like sodium, has only a single valence electron, it is not metallic. Instead, hydrogen atoms form covalently bonded diatomic molecules.

Why does hydrogen behave so differently from sodium and the other metals? Formation of a localized covalent bond between two hydrogen atoms results in exposure of the peripheral regions of the molecule to the core charge of the hydrogen atoms.

In this case, because there are no core electrons, the H atom core charge is the nuclear charge of 1+. In the case of sodium and other metals, this situation leads to aggregation using the empty valence p orbitals.

However, hydrogen has no valence p orbitals. The single orbital in the n=1 shell is full; no further aggregation is possible. The formation of metallic bonds can occur only if empty valence orbitals are available.

Experiments have shown that if hydrogen is subjected to extremely high pressure, which forces the diatomic molecules to aggregate, it exhibits metallic characteristics.

In sodium, as in most metals, there are too few electrons to form normal two-electron bonds between all pairs of bonded atoms. To achieve stability, the valence electrons are spread over many nuclei, simultaneously bonding them all together.

This spreading of electrons is called *delocalization*, and in some ways is similar to the delocalization that occurs in resonance. In metals, however, delocalization is much more extensive, and it is best to think of the electrons as waves that extend over the entire dimensions of the metallic crystal. A schematic structure of metallic bonding.

The nuclei are represented as points, and the electrons as waves. This picture enables us to understand the properties of metals.

That the electrons are delocalized waves is consistent with the conductivity of metals, and with their flexibility and strength. Putting stress on metals moves the nuclei, but the waves adjust, maintaining bonding. Because the electrons are waves, they are everywhere in the crystal at once. They are mobile, rationalizing the electrical and thermal conductivity of metals.

Finally, the pooling of valence orbitals by aggregated sodium atoms leads to a large number of closely spaced orbitals in the metallic crystal.

Thus the electrons absorb and reemit photons of all energies in the visible region of the spectrum, giving metals their highly reflective appearance.

There are so few bonding electrons in metallic sodium that only about 1/8 of an electron pair bond holds each pair of adjacent sodium atoms together. It is little surprise, then, that sodium is very soft (weak bonding), very reactive (deficiency of bonding electrons), with low density, and low melting and boiling points.

Example. In magnesium, how many metallic bonds form per Mg atom?

Solution. The electron configuration of the magnesium atom is $1s^2\ 2s^2\ 2p^6\ 3s^2$. We assume that metallic bonds are formed using the valence electrons only. Because each Mg atom contributes 2 valence electrons to the metallic bonding, and because 2 electrons constitute one bond in Lewis terms, there is one bond per Mg atom in the metallic lattice.

INTERMOLECULAR FORCES

Having described the forces that operate within molecules (including ionic crystals and metals), we are ready to discuss the interactions that occur between species (atoms, molecules, and ions) that are not covalently bonded to each other. These are called *intermolecular forces*; they can originate in a number of ways.

Dipole-Dipole Forces

Intermolecular forces that operate between neutral molecules having molecular dipole moments are called dipole-dipole forces. The structure of a molecule may cause its centers of positive and negative charge to be displaced from one another, giving the molecule a dipole moment. The dipole moments of two neighboring molecules tend to align with the + end of one dipole near the–end of the other, so that forces of attraction between them are maximized.

The maximum force of attraction between two dipoles, m_1 and m_2, separated by a distance r is given by equation.

$$F = \mu_1 \times \mu_2/\Omega \times r^4$$

Equation is Coulomb's Law, in disguised form. It is necessary to raise the distance of separation to the 4th power in the denominator to cancel the distance units that appear in both dipole moments in the numerator. Physically, the 4th power dependence means that the attraction between dipoles falls off much more rapidly with distance than would the attraction between isolated charges.

This is because the + and–ends of dipole 2 are almost equidistant from dipole 1 when the dipoles are separated by a distance that is much greater than the dipole length. The force of attraction between the + end of dipole 1 and the–end of dipole 2 is almost cancelled by the force of repulsion between the two + ends. Dipole-dipole forces operate between molecules of water, and between molecules of the substances.

Such forces are obviously much weaker than those operating in ionic or covalent network solids, and give rise to potential wells having depths in the approximate range 5-20 kJ/mole. Many molecular substances with dipolar molecules exist as liquids at ambient temperature, and have relatively low boiling points. In particular, many organic (carbon-containing) compounds are of this type.

Example. Which substances experience dipole-dipole intermolecular forces?

$$SiF_4, CHCl_3, CO_2, SO_2$$

Solution. We can draw the following conclusions about the molecules of these substances.

SiF_4	tetrahedral	Si-F bonds are polar, but no molecular dipole; bond dipoles cancel
CO_2	linear	C-O bonds are polar, but no molecular dipole; bond dipoles cancel
SO_2	bent	S-O bonds are polar and do not cancel. Sulfur lone pair dipole only partially offsets net bond dipole.
$CHCl_3$	tetrahedral	C-H and C-Cl bonds are polar and do not cancel

Hydrogen Bonding.

Hydrogen bonding (H-bonding) is a special kind of dipole-dipole force that occurs when a hydrogen atom is bonded to one of the very electronegative atoms, F, O, or N. The H-F, H-O, and H-N bonds are very polar, because the electronegative atom draws the bonding electron pair strongly to itself.

This leaves the hydrogen nucleus exposed. The resulting bond dipole is substantial. The positive end of this dipole, consisting essentially of an exposed hydrogen nucleus, has a high charge density because the proton is so small. Consequently, it strongly attracts the negative end of a similar bond dipole in another molecule to give a structure.

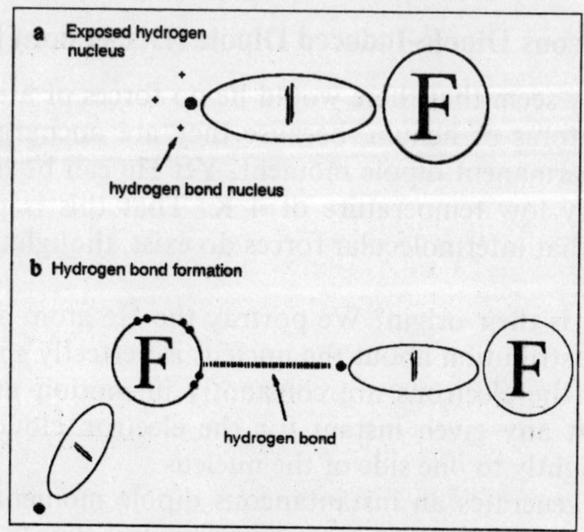

Fig. Hydrogen Bonding

The hydrogen atom is sandwiched between the two electronegative atoms in a linear arrangement. Hydrogen bond forces cause potential wells of depth in the range 5-50 kJ/mole. Note that these are on average deeper than those resulting from ordinary dipole-dipole forces. The intermolecular forces between water molecules are of the hydrogen bonding type.

They are responsible for the abnormally high boiling point of water, which, with its small molecular weight, would otherwise be a gas at room temperature. Compare the boiling point of water (100 °C) with that of methane (-162 °C), which has a similar molecular weight, but lacks hydrogen bonding and ordinary dipole-dipole forces. Hydrogen bonding is responsible for the formation of genes in the DNA molecule; for the helical structure of proteins; and for the incredible strength of Kevlar, a DuPont polymer used for canoe hulls and bullet-proof vests.

Example. In which substances do hydrogen bonding forces operate between molecules?

$$CHCl_3, CH_3CH_2OH, HNO_3, PH_3$$

Solution. Hydrogen is bonded to one of the very electronegative atoms in CH_3CH_2OH and HNO_3. Hydrogen bonding should occur in both of these substances.

Instantaneous Dipole-Induced Dipole (Dispersion) Forces

It may seem that there would be no forces of attraction between atoms of helium, because they are uncharged and have no permanent dipole moment. Yet He can be liquified at the very low temperature of 4 K. That this is possible indicates that intermolecular forces do exist, though they are very weak.

What is their origin? We portray the He atom as if the electron distribution about the nucleus is perfectly spherical. However, the electrons are constantly in motion and it is possible at any given instant for the electron cloud to be skewed slightly to one side of the nucleus.

This generates an instantaneous dipole moment in the He atom. During its transitory existence, this dipole induces a dipole in a neighboring atom and the two instantaneous dipoles attract. Forces between transitory dipoles are called instantaneous dipole-induced dipole forces, or alternately, London dispersion forces after the scientist who first proposed them. We will refer to them as *dispersion forces*.

It is these forces that are responsible for the liquefaction and/or solidification of a number of substances whose molecules do not possess permanent dipole moments. For example, dry ice is solid CO_2. The CO_2 molecule is linear and, though it has non-zero bond dipoles, has no overall dipole moment. Similarly, CCl_4, with perfectly tetrahedral molecules and no net dipole moment, is a liquid at room temperature and up to 78 °C. And iodine, which consist of I_2 molecules, is a solid at room temperature.

Dispersion forces depend on two features of molecular structure. First, they increase in magnitude with the size and distortability (usually called the polarizability) of the electron clouds of the interacting particles. Size and polarizability increase as molecular weight increases.

It follows that dispersion forces increase as molar mass increases. For substances of large atomic or molecular mass, dispersion forces are strong enough that the substances are solid or liquid at room temperature. We saw this above for CCl_4 and I_2. A striking example is provided by the diatomic halogens of group 17, which progress from F_2, a gas at room temperature and 1 atm pressure, to Cl_2, also a gas, to Br_2, a liquid, and finally to solid I_2. Only dispersion forces are operative in this series of substances. Second, dispersion forces depend upon molecular shape via the surface area over which two molecules can be in contact.

The larger the surface area of contact, the stronger the dispersion forces. Molecules that are roughly spherical in shape are able to contact each other only minimally. In contrast, molecules that are planar or linear in shape can maintain a large surface area of contact, with correspondingly larger dispersion forces. The classic example of the effect of molecular symmetry (shape) on the magnitude of dispersion forces is provided by the series of isomeric pentanes (C_5H_{12}).

Neopentane is highly symmetrical and nearly spherical in shape. Intermolecular forces are relatively small and the boiling point is correspondingly low. Pentane is linear, so two pentane molecules are able to contact each other along the entirety of their length.

Intermolecular forces are relatively large, and the boiling point is correspondingly high. Isopentane is intermediate structurally, and the least symmetrical of the three isomers. Intermolecular forces are less than in pentane, but greater than in neopentane.

The boiling point is intermediate. Graphite, the most stable form of elemental carbon at ordinary temperature and pressure, has the structure. Each carbon atom is bonded to three others in trigonal planar fashion, using sp^2 hybrid orbitals. The unhybridized p orbitals are used to form a network of pi bonds that extends over the entire array of bonded carbons, producing a large planar sheet of carbon atoms. These sheets are then stacked one atop the other to produce the three dimensional structure of graphite. Interestingly, there are no covalent bonds holding one sheet to the next.

The sheets attract one another via dispersion forces, which are very strong because of the large surface area of contact.

It is important to realise that dispersion forces operate between all molecules, whether or not other forces also operate. Molecules of chloroform, $CHCl_3$, are atttracted by a combination of dipole-dipole and dispersion forces. The magnitude of dispersion forces varies with molar mass; however, they are generally weaker than dipole-dipole forces, giving potential wells in the range 0.1 to 5 kJ/mole.

Example. Arrange the following non-polar molecules in order of increasing melting point.

$$SiF_4, CS_2, CI_4, GeCl_4$$

Solution. None of these molecules is expected to have a dipole moment. Only dispersion forces will operate, and these increase with increasing molecular mass. The molar masses of these substances follow:

Substance	*MM*
SiF_4	104.077
CS_2	76.131
CI_4	519.631
$GeCl_4$	214.402

The intermolecular forces, and the melting points, should increase in the following order:

$$CS_2 < SiF_4 < GeCl_4 < CI_4$$

The experimentally determined melting points are -110.8, -90, -49.5, and 171 °C, respectively.

Example. Molar masses, dipole moments, and boiling points for several covalently-bonded substances are given below. Rationalize the observed boiling points in terms of intermolecular forces.

Substance	*MM*	μ	*Boiling Point, K*
He	4.003	0	4
H_2	2.016	0	20
N_2	28.014	0	77
CO	28.010	0.33	83
HCl	36.461	3.60	188
HBr	80.912	2.67	206
SO_2	64.058	5.42	263
HF	20.006	6.37	292
CCl_4	153.823	0	350
H_2O	18.015	6.17	373

Solution. The substances fall into two categories, those with dipole moments and those without.

Only dispersion forces operate in substances without dipole moments. Selecting from the table those substances with no dipole moment, shown below, we see that, with the exception of hydrogen, boiling point increases with increasing MM, as expected from our discussion of dispersion forces above.

The dipole moment of CO is so small that it seems reasonable to include it in this list.

Its boiling point is only slightly larger than that of N_2, due to weak dipole-dipole forces in addition to the dispersion forces.

Substance	*MM*	*μ*	*Boiling Point, K*
He	4.003	0	4
H_2	2.016	0	20
N_2	28.014	0	77
CCl_4	153.823	0	350

Substances having substantial dipole moments are listed below, now in order of increasing dipole moment.

Substance	*MM*	*μ*	*Boiling Point, K*
HBr	80.912	2.67	206
HCl	36.461	3.60	188
SO_2	64.058	5.42	263
H_2O	18.015	6.17	373
HF	20.006	6.37	292

We see only a rough correlation of boiling point with dipole moment: HBr and HCl have boiling points on the low end of the scale; SO_2 is intermediate; and H_2O and HF have high-end boiling points.

However, the boiling points of HBr and HCl are inverted based on the size of the dipole moment. HCl, with the larger dipole moment, has the lowest boiling point in the group of five. Here is clear evidence that dispersion forces operate in addition to dipole-dipole forces.

Dispersion forces are substantially larger for HBr (MM 80.9) than for HCl (MM 36.5). This is sufficient to raise the boiling point for HBr above that for HCl.

The remaining 3 substances do not yield to the same argument, however. The MM of SO_2 is more than three times larger than those of HF and H_2O, and its dipole moment is only somewhat less than that for water.

Yet the boiling points of HF and particularly of H_2O are strikingly higher than that for SO_2. Here we have evidence of the potency of hydrogen bonding intermolecular forces, which operate in both HF and H_2O.

Despite its small MM, HF boils 30° higher than SO_2 because hydrogen bonding causes the potential well of the liquid to be relatively deep.

The boiling point for water is about 80° higher than that for HF because each water molecule can engage in up to four hydrogen bonds (one for each H atom, and one for each lone pair on the O atom).

Mixed Forces in Solutions.

All of the forces that can operate at the molecular level within a single pure substance are discussed above. It is possible to have what we might call hydrid forces that operate when pure substances are mixed to form solutions. When an ionic solid such as sodium chloride dissolves in water, ion-dipole forces are responsible. The maximum magnitude of ion-dipole forces is given by a form of Coulomb's Law that is intermediate between equations.

$$F = q \times \mu/\Omega \times r^3$$

Here q is the charge on the ion, m is the dipole moment of the solvent, and r is the distance between the ion and the centre of the solvent dipole.

Example. Consider the following three situations:

- A charge of 1+ separated by 2.0×10^{-10} m from a charge of 1- (this might be two ions in a crystal);
- A charge of 1+ separated by 2.0×10^{-10} m from a dipole of magnitude 1.6×10^{-29} C-m (a dipole of this magnitude would result from charges of 1+ and 1- separated by a distance of 1.0×10^{-10} m).
- Two dipoles, each of magnitude 1.6×10^{-29} C-m, separated by 2.0×10^{-10} m.

Compare the magnitudes of the attractive forces operating in these situations. This will provide an idea of the relative magnitudes of ion-ion, ion-dipole, and dipole-dipole forces. Then, compare the forces when the distance of separation of the interacting entities is increased by a factor of 4 to 8.0×10^{-10} m.

Solution. We apply equations:

Ion-ion: $F = (1.6 \times 10^{-19}\ C)^2/\Omega \times (2.0 \times 10^{-10})^2 = 6.4 \times 10^{-19}/\Omega$ Newtons

Ion-dipole $F = (1.6 \times 10^{-19})(1.6 \times 10^{-19})/\Omega \times (2.0 \times 10^{-10})^3 = 3.2 \times 10^{-19}/\Omega$ Newtons

Dipole-dipole $F = (1.6 \times 10^{-29})^2/\Omega \times (2.0 \times 10^{-10})^4 = 1.6 \times 10^{-19}/\Omega$ Newtons

The ion-ion force is stronger by a factor of 4 than the dipole-dipole force. The ion-dipole force is intermediate.

When the separation distance is increased by a factor of 4, the magnitudes of all forces decrease because they depend inversely on a power of the distance. The ion-ion force is reduced by a factor of 4^2; the ion-dipole force by a factor of 4^3; and the dipole-dipole force by a factor of 4^4:

Ion-ion: $F = 4\times10^{-20}/\Omega$ NtIon-dipole $F = 0.5\times10^{-20}/\Omega$ NtDipole-dipole $F = 0.063\times10^{-20}/\Omega$ Nt

The forces involving dipoles clearly decrease much more rapidly with increasing distance of separation than do ion-ion forces.

Whether or not two liquids are mutually soluble (miscible) depends on whether the mixed intermolecular forces can compete with the forces operating within each pure liquid alone. The intermolecular forces in pure water are primarily hydrogen bonding (dipole-dipole) forces, superimposed on weak dispersion forces.

Those in pure acetone are dipole-dipole forces superimposed on dispersion forces that are stronger than the dispersion forces in water. When acetone and water are mixed, they readily dissolve in one another in all proportions, apparently because the dipole-dipole forces between water and acetone molecules are of strengths comparable to those operating in the two separate liquids. On the other hand, only dispersion forces operate within pure carbon tetrachloride, because the CCl_4 molecule has a molecular dipole moment of zero. Because the molecules are spherical, giving small surface area of contact, these forces are fairly weak.

Consequently, carbon tetrachloride and water are not mutually soluble. They are said to be *immiscible*. When CCl_4 and water are mixed, the two liquids segregate into layers, with the denser liquid, chloroform, on the bottom. The liquid-liquid interface (boundary) is clearly visible.

Finally, a non-polar solid like I_2 dissolves readily in a non-polar liquid like CCl_4, because the mixed dispersion forces are comparable in magnitude to the dispersion forces in pure CCl_4 and pure I_2. However, the solubility of I_2 in water is essentially zero, because mixed forces cannot compete with the strong dipole-dipole forces in water. Water molecules are unwilling to separate in order to make room for the non-polar I_2 molecules. In this way, we can extend our intermolecular force ideas to provide an understanding of the formation of mixtures.

Summary of Mixed-Force Concepts

- Substance A will dissolve in substance B if the forces operating between A and B molecules are of similar type (dipolar or dispersion) and strength to those operating between A molecules and between B molecules. This statement is often sloganized: "Like dissolves like."
- Substance A will not dissolve in substance B if the forces operating between A and B molecules are of different types (one dipolar, one dispersion) (and therefore usually different strengths) than those operating between A molecules and between B molecules. This can also be sloganized: "Like does not dissolve unlike."

Example. For each pair, predict whether substance A will dissolve in liquid B, and explain your reasoning.

Substance A	*Liquid B*
C_2H_5OH	CH_3OH
Br_2(l)	C_5H_{12}
Br_2(l)	H_2O
C_2H_5OH	H_2O
C_2H_5OH	$CHCl_3$
NaCl	$CHCl_3$
NaCl	CH_3CH_2OH
H_2O	CH_2Cl_2

Solution. C_2H_5OH (ethanol) and CH_3OH (methanol) are both liquids due to fairly strong H-bonding intermolecular forces. Ethanol is expected to dissolve in methanol.

Br_2(l) and C_5H_{12} (pentane) are both non-polar liquids with dispersion forces operating between molecules. Thus they are "like" and we expect bromine to dissolve in pentane.

Br_2 and H_2O are "unlike." Bromine will not dissolve readily in water.

Both liquids exert H-bonding forces at the molecular level. Ethanol is expected to dissolve in water.

This is not obvious. At first glance, we would call ethanol and chloroform "unlike" because H-bonding forces operate in ethanol, whereas dispersion forces operate in chloroform. However, in addition to its polar -OH group, ethanol has a non-polar hydrocarbon portion that is capable of exerting dispersion forces on molecules of chloroform. These two liquids are therefore miscible.

Ionic NaCl will certainly not dissolve in non-polar chloroform.

Because ethanol is polar at the -OH end, we might expect NaCl to dissolve to some extent in ethanol, but not to the same extent as in water. This is indeed found. Polar water will not mix with non-polar dichloromethane.

Example. Describe what the system will look like during and following each process:

- 50 mL of distilled water is added to an Erlenmeyer flask.
- 0.10 g of $I_2(s)$ is added to the flask, and the flask is stoppered and vigorously shaken.
- 50 mL of CCl_4 is added to the Erlenmeyer flask, and the flask is stoppered and shaken.

Solution.

- The clear, colorless liquid will occupy the bottom of the flask.
- I_2, and intensely violet solid, is not expected to be soluble in water because of incompatible intermolecular forces. Because I_2 is more dense than water, it will sink to the bottom of the flask. The water will remain clear and colorless.
- CCl_4 is a non-polar liquid in which the intermolecular forces are of the dispersion type. It is expected to be immiscible with water. Because it is more dense than water, it will sink to the bottom of the flask, and the water will float on it with a distinct liquid-liquid

boundary. CCl_4 and I_2 have compatible intermolecular forces, so I_2 is expected to dissolve readily in carbon tetrachloride to give a purple solution. The flask will thus contain a layer of purple liquid under a layer of colorless liquid.

Distribution of a solute Between Two Immiscible Liquids in Contact

When two immiscible liquids, such as water and chloroform ($CHCl_3$), are placed in contact, they will mutually exclude each other, forming two liquid layers. The less dense liquid will float on the denser one, and a boundary surface will be clearly visible.

It is interesting to think about what will happen if a third substance, solid or liquid, is added to the system consisting of two immiscible liquids, A and B.

If the third substance, C, is soluble in one of the liquids (say A) and insoluble in the other (B), it will dissolve exclusively in the A layer, as in example 6-14. More interesting is the situation in which the C is soluble in both liquids to at least some extent. In this case, we find that the substance C *partitions*, or *distributes*, itself between the two liquid layers. We further find that the concentration of the third substance is higher in the liquid with which its intermolecular forces are most compatible. The ratio of the concentration of C in liquid A to that in liquid B is found to attain a constant value K_d, defined as in equation:

$$K_d = [C]_A/[C]_B$$

This constant ratio is established no matter how the 3-component mixture is prepared. Three such ways follow:

- A and B are first brought into contact. C is then added, and the container is shaken.
- C is first dissolved in A. A (containing C) and B are then brought into contact, and the container is shaken.
- C is first dissolved in B. A and B (containing C) are then brought into contact, and the container is shaken.

In addition, the ratio is reestablished if the mixture is altered in any of the following ways:

- More A is added.
- More B is added.
- More C is added.

We can understand this behaviour in terms of a simple layers of A and B in contact at a boundary, and molecules of C initially dissolved in liquid A. The boundary between the two liquids provides a "doorway" from one to the other through which C molecules, which have reasonable interactions with both A and B molecules, can pass.

Thus molecules of C will tend to move across the boundary from liquid A, where they are initially dissolved, into liquid B, in which there are initially no molecules of C. In other words, they begin to distribute themselves between the two liquids. After some time, enough C molecules will have passed into liquid B that the rates of passage of C from A to B and back will be the same. Once this situation is attained, there will be no further change in the concentration of C in the two layers. In other words, the constant ratio in equation will be attained. From the magnitude of K_d, we may easily deduce which liquid C prefers. If $K_d < 1$, C prefers B; if > 1, C prefers A.

Distribution of a substance between two liquid phases is an example of dynamic equilibrium, a situation in which two opposing processes (here, movement of C both ways across the boundary) occur at equal rates.

A Nod to Other Classification Systems

Classification of the various types of forces as "inter-" or "intramolecular" is not absolute. Thus you may see the forces classified as follows:

- Covalent Bonds (in molecular and network covalent substances)
- Ionic Bonds
- Metallic Bonds
- Interunit forces
  - Ion-dipole

- Dipole-dipole
- Hydrogen bonds
- Dispersion forces

In this classification scheme, ionic and metallic bonds are categorized separately rather than as intermolecular forces, because there are no discreet molecules in ionic compounds or metals.

Another classification system has recently become popular and is experiencing increasing use. This goes as follows.

- Covalent Interactions (includes covalent bonds in molecular and network covalent substances)
- Non-covalent Interactions
  - Ionic bonds
  - Metallic bonds
  - Metal-ligand interactions
  - ion-dipole forces
  - dipole-dipole forces
  - hydrogen bonds
  - dispersion forces

Here there are only 2 major categories: covalent, and not covalent! It is important to realise that the distinction between covalent bonding and some of the non-covalent interactions is blurred. For example, many people believe that metal-ligand interactions and even hydrogen bonding are weak covalent interactions.

The bottom line is that it is not particularly important how the classification is done, as long as the various types of interactions are understood. Pick your own favourite classification system, and use it!

The Interplay Between Molecular Kinetic and Potential Energies

The concept of temperature is of profound significance in science. We take it for granted in our daily lives, without thinking much about what it is. Temperature is the manifestation of molecular kinetic energy. The steam rising from a pan of boiling water feels hot because the molecules of water vapour are moving rapidly. Their collisions with the skin transfer kinetic energy, which we interpret as heat. The relationship between temperature and the average translational kinetic energy of an atomic, ionic, or molecular unit is very simple.

$$\text{Average } KE_{molecule} = 3/2\ kT$$

T is the absolute (Kelvin) temperature, and k is a fundamental constant of nature called Boltzmann's constant. Its value is 1.381×10^{-23} J/K. By $KE_{molecule}$, we signify the kinetic energy of any microscopic unit of matter. This can be anything from an electron to a DNA molecule. Keep in mind in using equation that the kinetic energy referred to is the translational kinetic energy—the energy of motion of the particle as a whole through space.

Translational kinetic energy is often called thermal energy.

Example. Calculate the thermal energy of one mole of nitrogen gas at room temperature, 293 K.

Solution. Calculate the kinetic energy of an average molecule; then multiply by Avogadro's Number.

$$\text{Average } KE_{molecule} = 3kT/2 = 3(1.381 \times 10^{-23}\ \text{J/K})(293\ \text{K})/2$$
$$= 6.07 \times 10^{-21}\ \text{J/molecule}$$

$$KE_{mole} = N_o KE_{molecule} = 3N_o kT/2$$
$$= 6.07 \times 10^{-21}\ \text{J/molecule} \times 6.023 \times 10^{23}\ \text{molecules/mole}$$
$$= 3.66 \times 10^{3}\ \text{J/mole}$$

The product of Boltzmann's constant and N_o is a constant, R, which occurs in the ideal gas law. Its value is 8.314 J/K-mole. Expressed in terms of R, the kinetic energy per mole of gas is 3RT/2.

We have seen that molecular forces promote aggregation by lowering molecular potential energy. Kinetic energy counteracts the effects of intermolecular forces. These ideas lead to a very useful generalization:

Potential energy favors aggregation; kinetic energy opposes aggregation.

The stronger the molecular forces, the more rapidly

atoms, molecules, or ions must move in order to overcome them. Even the strongest forces, however, may be overcome by the vigour of molecular motion if the temperature is high enough. At a temperature that gives a kinetic energy comparable in magnitude to the depth of the potential well, molecules begin to escape the potential well, and the substance undergoes a phase change (that is, sublimes, melts, or boils).

A phase change may therefore by viewed as an exchange of KE for PE. If the potential well is shallow, as in liquid helium (where the only forces are the very weak dispersion type) the required kinetic energy is achieved at very low temperature (4 K). If the potential well is deep, as in the ionic compound MgO (where the forces are the strong ion-ion type), very high temperature is required for particles to escape (3125 K). Table shows the parallel between the potential well depth and the temperature of melting for a number of substances.

Table. Melting Temperature and Potential Well Depth

Substance	*Potential Well Depth, J/mole*	*Melting T, K*
He	0.105	3.5
H_2	1.04	14.0
Xe	14.7	161
N_2	6.30	63.2
Cl_2	26.8	172.1
I_2	56.3	387
H_2O	46.7	273
CO_2	33.6	217
NaCl	771	1074

Try plotting melting point against potential well depth to see how good the correlation is.

NON-IDEALITY OF GASES

The postulates of KMT define an ideal gas. An ideal gas obeys the ideal gas law under all conditions of T and P. There is in fact no real gas that satisfies this definition. The ideal gas is a concept that we can think about but which does not actually exist.

Consider a sample of N_2 gas at 25 °C and 1 atm pressure. Imagine measuring the volume of the gas periodically as we cool it at constant pressure. If N_2 were ideal, it would follow Charles' Law exactly even to absolute zero. A plot of V versus T for ideal N_2 would follow curve A in Figure.

The actual V-T plot for N_2 follows curve A to a temperature of about 100 K. At this point the plot starts to curve down (curve B). At 77 K the volume sharply decreases while the temperature remains constant, and liquid appears in the container.

When all gas has converted to liquid, further decrease in T has little effect on the volume occupied by the liquid until T = 63 K. At this point, N_2 solidifies with a slight decrease in volume. Further cooling causes negligible change in V.

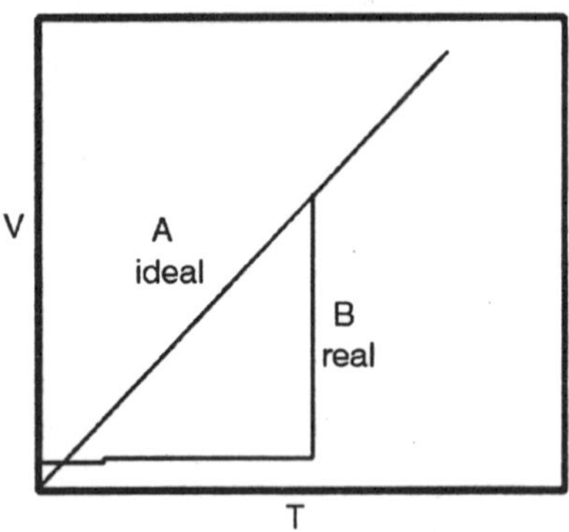

Because all real substances liquify at some T, no real substance can obey the ideal gas law under all P,T conditions. Deviation of real gases from ideal behaviour is shown conveniently in terms of the compressibility factor, Z

$$: Z = PV/nRT$$

Z is plotted against P in Figure. For an ideal gas, Z = 1 at all pressures (the horizontal dashed line). Two deviations from ideality are commonly observed. In one type, Z always exceeds 1. H_2 is an example. In the second type, Z < 1 at moderate pressure but becomes > 1 at high pressure. N_2 is an example. Most gases show behaviour similar to that of N_2. Real gases deviate from ideality because they violate two postulates of the KMT.

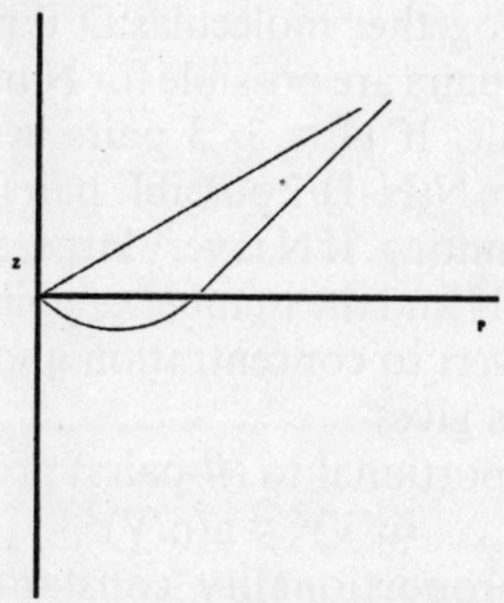

Fig. The Compressibility Factor, Z

First, real molecules exert attractive forces on one another. The forces cause the molecules to temporarily stick together in pairs when they collide, reducing the effective number of free particles in the gas. This is shown in Figure. Although there are 7 gas molecules in the container, 4 of them are temporarily stuck together in pairs, reducing the effective number of independent gas particles to 5. Sticking together reduces the pressure below the ideal value because there are fewer collisions with the container walls per unit time.

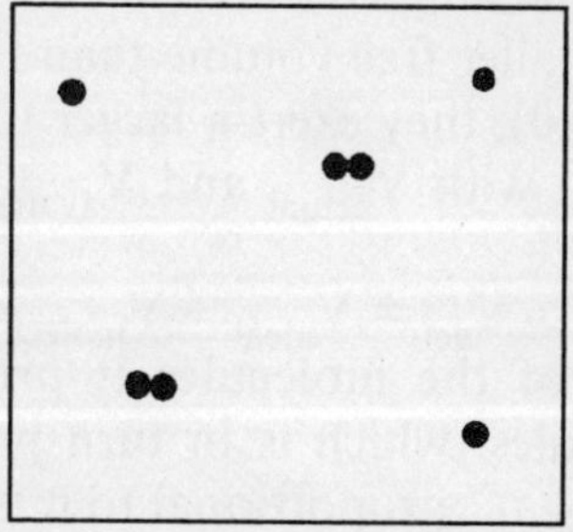

Fig. Non-Ideality of Gases

$$\text{Thus, } P_{act} = P_{ideal} - DP$$

Here P_{act} is the actual pressure of the gas, P_{ideal} is the pressure it would exert in the absence of intermolecular forces, and D P is the pressure reduction. If there are n moles of gas in a container of volume V, the pressure reduction, D P, is proportional to $(n/V)^2$. The following argument demonstrates this. First, we assume that D P is proportional to the number of pairs of stuck-together molecules:D P proportional to (# pairs). How many pairs are possible for N molecules? If N = 2, there is only 1 pair. If N = 3, 3 pairs are possible. For N molecules, there are N(N-1)/2 possible pairs, where division by 2 avoids double counting. If N is very large, as in a macroscopic gas sample, N-1 = N and the number of pairs is $N^2/2$. Dividing by volume to convert to concentration and by N_o to convert molecules to moles gives

$$D\,P \text{ proportional to (\# pairs) proportional to } (n/V)^2 = a(n/V)^2$$

where "a" is a proportionality constant. Substitution in equation gives $P_{ideal} = P_{act} + a(n/V)^2$

Second, real molecules are not points, but instead occupy finite volume. Thus a small fraction of available container space is occupied by the molecules themselves, and is not available to accommodate motion of the molecules. The total container volume is the sum of the volume of the molecules and the free volume available for motion:

$$V_{container} = V_{available} + V_{molecules}$$

The volume of the molecules is constant for a given gas, and depends on molecular size. As $V_{container}$ is made smaller, $V_{available}$ becomes a smaller fraction of $V_{container}$. Since the molecules have smaller free volume than they would if ideal (when $V_{molecules} = 0$), they exert a larger than ideal pressure. Equating $V_{container}$ with V_{actual} and $V_{available}$ with V_{ideal}, we obtain

$$V_{act} = V_{ideal} + V_{molecules}$$

The volume of the molecules is proportional to the number of molecules, which is in turn proportional to the moles of gas: $V_{molecules}$ proportional to n = bn

where b is a proportionality constant roughly equal to the volume per mole of substance. (The volume of the molecules can be determined by crowding the molecules together—i.e., by liquifying the substance. The molar volume of the liquid phase gives an estimate of "b". The molar volume of liquid is the product of the MW and the reciprocal of density:

b(vol/mole) = MM(mass/mole)×1/r (vol/mass) = MM/r (vol/mole)

Using this approach, b should be 18/1.0 = 18 mL/mole for water.) Substituting equation rearranging for V_{ideal} gives

$$V_{ideal} = V_{act} - bn$$

Finally we substitute equations into the ideal gas equation, $P_{ideal}V_{ideal}$ = nRT to obtain equation:

$$(P + an^2/V^2)(V-nb) = nRT$$

This equation was first proposed by van der Waals in 1873 to explain deviations of gases from ideal behaviour. Its success earned him the Nobel Prize for Physics in 1910.

In real gases the effects of non-zero molecular volume and intermolecular forces oppose. The former causes pressure to be higher than ideal, and the latter causes it to be lower than ideal. For N_2 and many other gases having $Z < 1$ at moderate pressure and $Z > 1$ at high pressure, first one effect, then the other, takes precedence. At low pressure (10-100 atm), the intermolecular forces dominate because the molecules are not yet compressed enough for the volume effect to be felt. This makes $P_{act} < P_{ideal}$, and $Z < 1$. As P becomes very large (> 100 atm), non-zero molecular volume becomes important, making $P_{act} > P_{ideal}$ and $Z > 1$.

For hydrogen, intermolecular forces are small and the volume effect dominates at all pressures. At low pressure (approximately 1 atm) and temperature much higher than the boiling point of the substance, both non-ideal effects are small and real gases approach ideal behaviour.

Values of the van der Waals constants a and b for several gases are in Table.

Table. Van der Waals Constants for Gases

Gas	*a, L^2-atm/mole*	*b, L/mole*	
Ar	1.35	0.0322	
Cl_2	6.49	0.0562	
CO_2	3.59	0.0427	
H_2	0.244	0.0266	
He	0.034	0.0237	
N_2	1.39	0.0391	
O_2	1.36	0.0318	

Example. Discuss the Van der Waals constants in Table in terms of molecular structure. ***Solution.*** The constant, a, is related to the magnitude of intermolecular forces; b is related to molecular volume. As a first guess, we might expect both of these to get larger with increasing molar mass of the gaseous substance. Let us predict that the constants will parallel molar mass. The substances and their constants are arranged in order of increasing MM below:

Gas	*MM*	*a, L^2-atm/mole*	*b, L/mole*
H_2	2.016	0.244	0.0266
He	4.003	0.034	0.0237
N_2	28.014	1.39	0.0391
O_2	31.998	1.36	0.0318
Ar	39.948	1.35	0.0322
CO_2	44.009	3.59	0.0427
Cl_2	70.906	6.49	0.0562

In a rough sense, our prediction is borne out. Both a and b trend upward as molar mass increases. However, we are wrong in the details. The value of a for He is a clear exception; and the constants for nitrogen, oxygen, and argon do not parallel the molar masses. Having predicted, somewhat but not entirely successfully, we now must rationalize (i.e., try to think of explanations for deviations from the trend).

First, we examine the hydrogen, helium pair. The difference between these is that hydrogen is molecular, and helium is atomic. Hydrogen molecules are highly symmetrical, but elongated from a perfect sphere.

The hydrogen molecules are more susceptible to dispersion forces than are the spherical, compact (high core charge) helium atoms. Consequently, forces (and the constant a) are larger for H_2, despite its lower molar mass. Incidentally, because there are 2 atoms per particle in hydrogen, the volume constant, b, is slightly larger for hydrogen.

Next, we consider nitrogen, oxygen, and argon, which have roughly similar molar masses. Again, Ar is atomic and spherical, whereas nitrogen and oxygen are molecular, with elongated electron distributions.

Despite their greater mass, the argon atoms are less susceptible to dispersion forces because they are spherically shaped. For the same reason, the volume constant for Ar is comparable to those for nitrogen and oxygen, despite the greater number of electrons.

Atomic size decreases left to right in a period; oxygen atoms are actually smaller than nitrogen atoms, despite their larger atomic weight. Thus the volume constant, b, is smaller for oxygen than for nitrogen. Oxygen is more electronegative than nitrogen, making it somewhat less distortable. Intermolecular forces, measured by a, are somewhat less in O_2.

SOME CONSEQUENCES OF INTERMOLECULAR FORCES

The operation of the types of forces between molecules, atoms, and ions has profound implications for the properties, function, and reactivity of chemical substances. By way of illustration, we discuss intermolecular forces in that most fundamental and essential of all molecules, water. The water molecule is bent, with a bond angle of about 104 °. This shape is understandable in terms of the VSEPR theory. A consequence of this shape is that the molecule is polar (i.e., has a molecular dipole).

The molecular dipole moment and the fact that water contains hydrogen bonded to the very electronegative oxygen atom leads to strong intermolecular forces of the hydrogen bonding (dipole-dipole) type.

A consequence of the existence of these forces is that the melting and boiling points of water are much higher than would be expected on the basis of molar mass (e.g., the boiling point of methane, with molar mass 16, is -162 °C). Thus water is a liquid or solid under temperature conditions found on the earth's surface.

As a liquid, it provides a medium for the genesis and sustenance of life. Another consequence of polarity is that water has excellent solvating properties: it dissolves substances as diverse in structure as salts, sugars, and huge protein molecules that function as enzymes. Due to the shape of the water molecule and the intermolecular hydrogen bonding, water adopts an open lattice structure when it freezes, in which each oxygen atom is covalently bonded to two hydrogen atoms and hydrogen bonded to two more via its two lone pairs of electrons.

A consequence of this open lattice is that ice is less dense than liquid water, and floats. Lakes therefore freeze top down, the layer of ice on top serving to insulate the unfrozen liquid below and thereby to protect aquatic life. If, like most substances, water were more dense as a solid, ice would sink, entire lakes would freeze, and life as we know it would not have arisen.

(Of course, the density inversion of water has negative consequences, too, with which we are all familiar: potholes in roads, erosion of rock formations, and the like.) Finally, the structure of the water molecule and the resulting intermolecular interactions are at the root of the ability of water to function as both an acid and a base. This aspect of the reactivity of water is crucial in many biochemical and geological processes.

Supplement: Chromatography

Separation and analysis of mixtures into pure components is an important problem in chemistry, with practical application in the forensics, cosmetics, food, liquor, petroleum, and paint industries. Modern chemists often use chromatography for such separations.

Table: Types of Chromatography

Type	*Stationary phase*	*Mobile phase*
Paper chromatography (GC)	Filter paper	Liquid, often water
Gas chromatography (GC)	Polar or non-polar liquid	Helium gas
High Performance liquid Chromatography (HPLC)	Solid	Liquid
Gel Permeation Chromatography (GPC)		
Thin Layer Chromatography (TLC)	Solid on glass or plastic plate	Liquid

Chromatography is a word used to encompass a range of techniques in which mixtures of pure substances are separated into the individual substances by using a mobile phase (usually a liquid or gas) to push the mixture along a stationary phase (usually a solid or liquid coated on a solid). Because the individual substances have different molecular structures, they interact differently with both the stationary and mobile phases, and consequently are "pushed" at different rates by the mobile phase. A number of chromatographic techniques are summarized in Table.

Thin Layer Chromatography

Thin-Layer Chromatography (TLC) is a simple and inexpensive technique that is often used to judge the purity of a synthesized compound or to indicate the extent of progress of a chemical reaction. In this technique, a small quantity of a solution of the mixture to be analyzed is deposited as a small spot on a TLC plate, which consists of a thin layer of silica gel (SiO_2) or alumina (Al_2O_3) coated on a glass or plastic sheet. The plate constitutes the stationary phase. The sheet is then placed in a chamber containing a small amount of solvent, which is the mobile phase.

The solvent gradually moves up the plate via capillary action, and it carries the deposited substances along with it at different rates. The desired result is that each component of the deposited mixture is moved a different distance up the plate by the solvent. The components then appear as a series of spots at different locations up the plate. Substances can be identified from their so-called R_f values.

The R_f value for a substance is the ratio of the distance that the substance travels to the distance that the solvent travels up the plate. For example, an R_f value of 0.5 means that the spot corresponding to the substance travels exactly half as far as the solvent travels along the plate.

The Process of TLC. Performing a TLC analysis consists of a number of steps: preparing a spotting capillary; spotting the TLC plate; developing the TLC plate; drying the plate; visualizing the substance spots, and measuring the R_f values. We consider these steps in turn.

Preparing a spotting capillary. Glass capillaries used for spotting TLC plates are commercially available. However, it is occasionally necessary to make your own capillaries. To accomplish this, light a Bunsen burner and adjust for a medium flame. Hold a melting point capillary in the flame until it just begins to soften, then quickly pull the two ends of the capillary in opposite directions.

The central, soft part of the glass will elongate and thin down to a capillary with very small diameter. Break the two pieces apart at the centre of the thin portion to obtain two TLC spotting capillaries.

Spotting the TLC plate. Obtain a silica gel TLC plate that is approximately 2 cm wide and 5 cm long. Using a pencil, draw a straight line parallel to the short dimension of the plate, about 1 cm from one end of the plate. This line will serve as a guide for placing the substance spots, and as a point from which to measure R_f values. Be sure to use pencil to draw this line, rather than pen, because inks will be moved by many developing solvents. Place the narrow end of one of your capillaries into a vial containing a solution of the substance to be analyzed, and allow the solution to rise in the capillary; this will happen spontaneously.

Once the capillary is loaded, hold it vertically just above the pencil line on the plate. Lower it until the narrow end of the capillary just touches the plate. You will observe that some of the solution leaves the capillary and deposits on the plate. Leave the capillary in contact with the plate only briefly so that the spot is no larger than 1 mm in diameter, then raise the capillary. Allow the solvent to completely evaporate from the spot. Then, if desired, make a second deposit on the same spot with the capillary. Again allow the solvent to completely evaporate.

Developing the TLC plate. Pour some of the desired developing solvent into a small wide-mouth glass or plastic bottle to a depth of about 4-5 mm. Using tweezers, pick up the TLC plate at the top, which is the end opposite where the pencil line is drawn. Place it carefully in the developing bottle so that it stands as nearly vertical as possible. It is important that you not allow the TLC slide to tilt too much when in the developing bottle.

If the slide is tilted, solvent will not advance uniformly

along the plate and development will not take place properly. Leave the slide in the chamber until solvent has advanced to within 8-10 mm of the top of the slide. Then use the tweezers to withdraw the slide from the chamber and quickly draw a pencil line marking the solvent front.

Drying the Plate. Place the plate flat on a clean dry surface and allow the solvent to completely evaporate. If the solvent is not highly volatile, this can be facilitated by placing the slide on a flat surface in an oven at a temperature of 50-60°C (higher temperatures will melt the plastic substrate material). When the plate is completely dry, it is ready for visualization.

Visualization of the TLC Plate. If the substances being separated are colored, the spots can be seen without any further effort. Using a pencil, draw a boundary around each spot that matches the shape of the spot. Many substances are colorless (white) and do not show up on the white silica gel unless steps are taken to make them visible. There are a number of techniques for doing this. *First* is the technique of iodination. The dry plate is placed in a chamber containing a few crystals of iodine. The iodine vapour in the chamber oxidizes the substances in the various spots, making them visible to the eye.

Once the spots are visible, they may be outlined with a pencil before the iodine coloration fades. *Second* is the ninhydrin technique, which is particularly effective for visualizing amino acid spots. In this method, a solution containing 0.2 per cent ninhydrin in ethanol is sprayed on the dry plate.

Alternately, the plate can be dipped in ninhydrin solution. In contact with an amino acid, ninhydrin displays a purple coloration that is easily seen. It usually takes a few minutes for this colour to develop, so after the plate is sprayed, it is allowed to sit for several minutes. Placing it in a 50°C oven will hasten the appearance of the purple colour. Once this appears, the spots may be circled in pencil to permanently mark their positions.

Third, one may use TLC plates that have been loaded

with a fluorescent substance that is uniformly distributed in the silica gel. Substances moving up the plate block this fluorescence at their locations. When the dried plate is viewed using a special UV light, the substance spots are visible for their lack of fluorescence on an otherwise uniformly fluorescing field. The spots may be outlined with pencil while being viewed in the light.

Measurement of R_f. The distance between the 2 horizontal pencil lines is the distance of solvent advance. The distance from the bottom pencil line to the centre of a substance spot is the distance of advance of the substance. The ratio of substance advance distance to solvent advance distance is the R_f value for the substance. The figure shows a developed plate for a mixture consisting of 4 components with Rf values of about 0.05, 0.2, 0.5, and 0.9

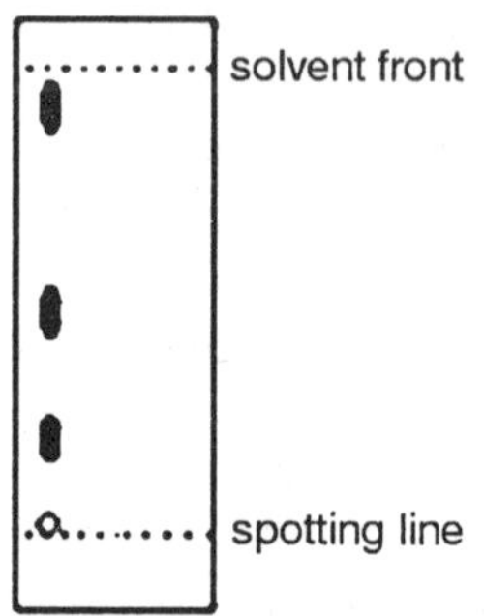

Finally, it is very important to be observant of detail in doing TLC. In addition to the R_f value for a substance, the shape of the spot produced by a particular developing solvent and the shade of colour produced by iodine or ninhydrin can be characteristic of the substance. Please note all of these things.

TLC of Inks. Before attempting to apply TLC to a real problem, it is advisable to learn and practice the technique by applying it to mixtures that are easily visualized and separated. Inks provide an ideal practice vehicle for TLC because they normally contain several colored components that separate nicely in common solvents such as ethanol, acetone, or chloroform. Spotting the plate is also easy: it may be done simply by making a VERY small mark on the plate with the tip of a pen, just above the pencil line drawn across the bottom edge of the plate. Inks are of different types and colors, of course.

Some are washable (water-soluble), others are permanent. Different types require different solvents for development. Common sources of ink are ball point pens, felt-tip markers, and roller ball pens. Bottled ink is still available for people using fountain pens. Particularly good ink sources are bottled inks by Parker, Sheaffer, and Mont Blanc; Sharpie marker inks (black, red, blue, orange, brown, yellow, green); Marks-a-Lot Stay sharp markers; and Sanford calligraphy pens.

TLC of Amino Acids. TLC of amino acids is more difficult than TLC of inks, because amino acids are colorless. Therefore, not only can you not monitor their progress up the plate, but you cannot see the spots with the naked eye once the plate is fully developed and dried. To see the spots, it is necessary to use either the ninhydrin or the black-light visualization techniques. Of course, the latter works only if you use fluorescent TLC plates. Until you see the spots, you will not know whether or not a chosen solvent system has been effective in moving an amino acid or in separating a mixture.

Therefore the process of finding an effective solvent system can be long and painstaking. As points of general information, amino acids are quite polar and tend to move on silica gel plates with polar solvents. They have R_f values close to 1 when water or concentrated ammonia is used as the developing solvent, probably because of their high solubility in water. Diluting a polar solvent with a less polar one results in smaller R_f values, roughly in proportion to the amount of less polar solvent used, Thus, alanine, glycine, threonine, and proline all have R_f values of around 0.60 when developed with a 50/50 mixture of water and n-propanol, and around 0.40 when developed with a 30/70 mixture of concentrated NH_3 and n-propanol.

When alanine, glycine, threonine, and proline are spotted side-by-side on a plate and developed with 70 per cent n-propanol/30 per cent conc NH_3, the following observations can be made:

Amino Acid	*Solvent*	*Spot Color after Iodination*	*Spot Color with Ninhydrin*	*R_f Value*	*Spot Shape*
alanine	30/70 conc NH_3/n-propanol	white on brown bkgrnd	purple		elongated oval
alanine	50/50 water/n-propanol	white on brown bkgrnd	purple		circle
glycine	30/70 conc NH_3/n-propanol	white on brown bkgrnd	pink		elongated oval
glycine	50/50 water/n-propanol	white on brown bkgrnd	pink		circle
threonine	30/70 conc NH_3/n-propanol	white on brown bkgrnd	purple		elongated oval
threonine	50/50 water/n-propanol	white on brown bkgrnd	purple		circle
proline	30/70 conc NH_3/n-propanol	dark brown on brown bkgrnd	yellow with pink border		elongated oval
proline	50/50 water/n-propanol	white on brown bkgrnd	yellow with pink border		circle

Problems in TLC.

Over-large Spots. Sample spots made using TLC capillaries should be no larger than 1-2 mm in diameter, because component spots in the developed plate will be no smaller than, and will usually be larger than, the size of the initial spot. If the initial spot is larger than 2 mm in diameter, then components with similar R_f values may not be resolved because their spots will be so large that they will overlap considerably and may appear to be one large spot. Small initial spots, on the other hand, maximize the potential of complete separation of components.

Uneven Advance of Solvent Front. A common problem in TLC is uneven advance of solvent along the plate. Instead of a straight line, the solvent front may appear to bow either up or down in the centre. Uneven advance of solvent leads to uneven advance of substance spots, and inaccurate R_f values result. A frequent cause of uneven solvent advance is the use of a developing chamber that does not have a flat bottom. Glass bottles usually have bottoms that curve upward from the edges to the centre.

If the bottom of the TLC plate is placed on this curved surface, the shape of the solvent advance line may mirror the shape of the container bottom. It is therefore important to use flat-bottomed developing tanks in TLC. A bowed solvent front may also result if too little developing solvent is placed in the chamber; if the plate is cut improperly, so that the sides are not exactly perpendicular to the bottom edge; and if the slide is allowed to deviate excessively from a vertical position in the chamber. Care in choosing and using a developing chamber is the best defence against curved solvent fronts.

Water is seldom used as a developing solvent because it has a tendency to produce a dramatically curved front. This may be due to its unusually high surface tension.

Streaking. Sometimes a substance will move along a TLC plate as a long streak, rather than as a single discrete spot. This is the result of spotting the plate with too much substance, more than the moving solvent can handle. The solvent moves as much as it can, but a substantial amount of substance is left behind. The substance is dragged along by the solvent leaving a trail of substance that may sometimes span the entire distance between the starting line and the solvent front. Streaking can be eliminated by systematically diluting the spotting solution until development and visualization show the substances moving as single spots, rather than elongated streaks.

Gas Chromatography. Gas chromatography is a bit more sophisticated than TLC. It is performed using a gas chromatograph, an instrument in which the mobile phase, usually helium gas, is passed through a *column* (a long narrow tube) containing the stationary phase. The essential design of a gas chromatograph is shown in the figure.

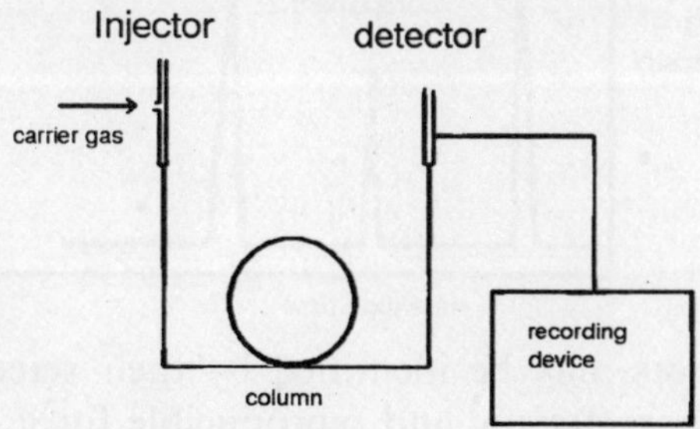

The three main sections of the gas chromatograph are the injector, the column, and the detector. Each of these may be separately thermostatted to any desired T between room T and about 250 °C. At the injector, a precise volume of the sample to be analyzed is added to the gas chromatograph using a syringe. A typical sample volume is 1-10 mL, where 1 mL = 1 mm^3). The injector must be hot enough to vaporize the sample. The sample vapour is picked up by the flow of helium carrier gas and swept onto the column.

The column is a long tube coated with a material (the stationary phase) which attracts molecules of the sample according to their structures. Molecules that are strongly attracted "stick" to (or are retained by) the column more tightly and take more time to pass through. Molecules that are weakly attracted take less time to pass thru. The total amount of time it takes after injection for a substance to pass through the column is called the retention time for the substance. The detector is a device that detects molecules of the sample and produces an electrical signal when they pass through. The electrical signal is sent to a recording device (often a computer) that shows a trace corresponding to the electrical signal produced by the detector.

The result of doing gas chromatograph of a sample is a chromatogram, a plot of detector signal output versus time elapsed since sample injection. It typically appears as shown in figure.

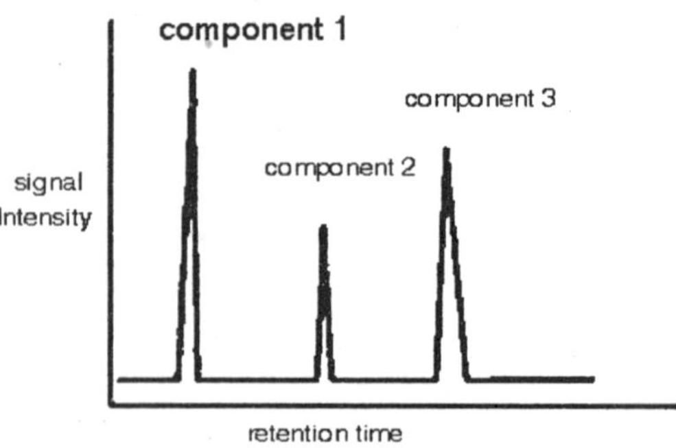

Substances may be identified by their retention times, which are characteristic and reproducible for a given set of injector, column, and detector temperatures, column stationary phase material, carrier gas flow rate, and so on. The signal intensities (measured as the area under a signal) are proportional to the molar amounts of the substances in the sample injected.

A number of simple and interesting problems can be solved using gas chromatography. An example is the determination of the active ingredient in fingernail polish remover. One might approach this problem by first listing some likely candidates, simple inexpensive non-toxic chemical substances that might do the job. Clearly water is not a good choice, because fingernail polish is untouched by it. However, some reasonable candidates are listed below:

- Acetone
- Ethanol
- Isopropanol
- Methanol
- Ethyl acetate

Having identified some likely candidates, one would proceed to use a gas chromatograph to determine their retention times on a particular column and with a particular set of injector, column, and detector temperatures. Some trial and error might be required to establish effective values for these temperatures.

Once the retention times for the candidates were known, it would be a simple matter to inject a sample of each of several commercial fingernail polish removers to the chromatograph, using the same set of temperatures and carrier gas flow rate.

If observed retention times for the removers matched any of the tested candidates, the problem would be solved. If not, other potential candidates would have to be identified and tested.

(P.S. The gas chromatographic experiment would show that commercial fingernail polish removers contain either acetone or ethyl acetate as the active ingredient. Ethyl acetate was chosen as a less toxic alternative for acetone when environmental concerns about the use of acetone were raised.)

Chapter 7

Solid and Liquid Phases

The vast majority of known substances are solids under ordinary conditions of pressure and temperature. An understanding of the structures of crystalline solids at the atomic/molecular level. Be forewarned that a study of such structures requires a sound basis in geometry and an ability to visualize in 3-dimensions.

Further, it is impossible to obtain an adequate understanding of the important structural ideas from text and 2-dimensional figures. Work with 3-dimensional models is essential. Understand it as best you can, then refine your understanding with models in the laboratory. A solid base for the understanding of two major classes of materials: metals, in which all atoms in the structure are the same; and ionic compounds, composed of cations of one substance and anions of another.

MACROSCOPIC PROPERTIES OF SOLIDS

Solids have a number of observable properties in common:

- They are rigid, maintaining fixed shape and volume independent of container.
- The molar volumes of solids are small.
- Solids are incompressible.
- Solids do not flow, and diffusion in solids is extremely slow.
- Many solids occur as crystals — well-defined shapes with structures suggesting an orderly internal arrangement of atoms, molecules, or ions.

To rationalize these properties we need a simple model of the solid phase. In designing such a model, we make use of two facts:

- Since the molar volume of a solid is much less than the molar volume of the gas phase of a substance, the molecules must be crowded together in the solid;
- The orderly shapes of crystals suggests an orderly arrangement of particles. A simple model, then, is that solids consist of tightly-packed orderly arrangements of particles. Figure shows a 2-dimensional representation of a solid.

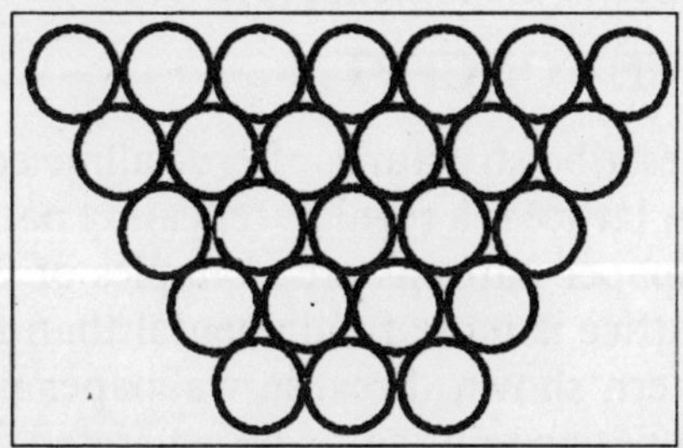

Fig. 2-dimensional representation of a solid

Note first how close together the particles are compared with the gas phase. This crowding allows very little free volume to the molecules, which strongly resist further compression. The solid is incompressible. It is also clear why diffusion is slow in solids, because each particle is locked in place by the surrounding ones. Only at the surface of the solid, where a particle has neighbors on only 3 sides, can diffusion occur.

The particles are in constant motion, just as in the gas phase. A molecule is constantly jiggling in place, bumping into neighbors that hold it in a cage. Just as for gases, the kinetic energies of the particles (atoms, molecules, or ions) of the solid are distributed according to the Maxwell-Boltzmann distribution, with the kinetic energy of an average particle having the value 3kT/2.

The major difference between the motion in gas and solid is that in the solid, a particle is hemmed in and undergoes many more collisions per unit time than in the gas. The movement of a particle in the solid is much like a vibration, because the molecule stays in the same location. If the temperature of the solid is increased, the particles vibrate faster, but are still not free to move about because intermolecular forces from the neighbors hold them in place. Finally, the arrangement of particles is very orderly, not at all chaotic like the gas.

Each particle is surrounded by a perfect hexagon of other particles; and particles are arranged in rows that remain straight over long distances. It is easy to imagine that this orderly arrangement at the molecular level is manifested in beautiful crystalline shapes at the macroscopic level.

THE LATTICE CONCEPT

We will describe structures of crystalline solids in terms of the lattice. A lattice is a regular, repeating pattern of points in space. Wallpaper patterns are examples of 2-dimensional lattices. The lattice is more fundamental than the particular wallpaper pattern shown, because wallpaper patterns using many different design units could be generated from the same lattice.

Figure shows a second small collection of points that might be considered as a unit cell. Movement of this collection in directions parallel to its sides does indeed reproduce the entire lattice.

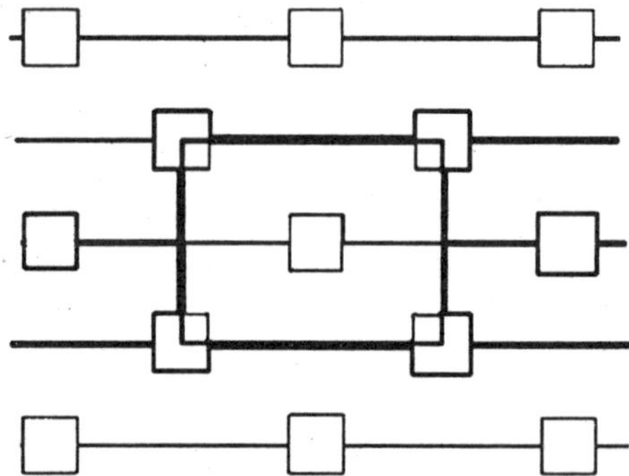

The key feature of the lattice is the arrangement of points. A small collection of lattice points that contains all of the spatial information in the lattice pattern is called a unit cell. The unit cell represents the lattice, because the whole lattice can be reproduced from the unit cell by moving (translating) the unit cell in directions parallel to its sides by distances equal to the lengths of its sides.

However, the unit cell in Figure is simpler than that in Figure because it involves fewer points — one point at each of its 4 corners. A unit cell having lattice points at its corners only is called a primitive unit cell. The unit cell in Figure is a non-primitive unit cell because it has an additional point at its centre. If it is possible to represent a lattice with a primitive unit cell, we will do so.

Crystalline solids are represented by 3-dimensional lattices, one example of which is the simple cubic lattice. This is a primitive unit cell because it has points only at the corners. Movement of the unit cell by the distance, d, in directions parallel to its 3 dimensions reproduces the entire simple cubic lattice. The simple cubic lattice is analogous to an orderly arrangement of building blocks stacked face to face in 3-dimensions.

It was shown in the 1800s by Bravais that there are only 7 shapes (called crystal types) that a unit cell can have and only 14 possible distributions of points (lattice types) within these 7 shapes. The 7 shapes, characterized by the relative magnitudes of the repeat distances in three directions and the angles between these directions, are summarized in Table.

Table. The Bravais Crystal Types

System	*Edge Lengths*	*Angles*
Cubic	a = b = c	all 90°
Tetragonal	a = b not equal to c	all 90°
Orthorhombic	a not equal to b not equal to c	all 90°
Monoclinic (once-angled)	a not equal to b not equal to c	$\alpha = \beta = 90°$, γ not equal to 90°
triclinic (thrice-angled)	a not equal to b not equal to c	α not equal to β not equal to γ not equal to 90°
Rhombohedral	a = b = c	= 90°
Hexagonal	a = b not equal to c	$\alpha = \beta = 90°$; $\gamma = 120°$

We will discuss cubic and hexagonal lattices in the context of a discussion of close-packing ideas.

CLOSE PACKING

What is the most efficient way of arranging spheres in 3-dimensional space? We take "efficient" to mean economical in terms of space. We begin by building one layer of spheres that are arranged as closely to one another as possible. Note that this arrangement is identical to that pictured earlier.

Spheres are lined up in straight rows, with the spheres in one row fitting snugly into the indentations in adjacent rows. Each sphere is surrounded by a perfect hexagon of spheres, all touching it. There are numerous gaps, or holes, in the layer, of triangular shape. For descriptive purposes here, we label these holes "u" if the triangle has a single vertex pointing up, or "d" if the triangle has vertex down.

We now add a single sphere of a second layer on top of the first. The added sphere rests in a low point of the first layer, corresponding with a triangular hole, in contact with 3 spheres of the first layer.

The 4 touching spheres lie at the vertices of a regular tetrahedron. There remains some unfilled space at the centre of this 4-sphere group that is called a tetrahedral hole. Each time we add a sphere to the second layer, we create a tetrahedral hole between the layers.

When additional spheres are added to the second layer, it turns out that they must all go in either "u" holes or "d" holes, but not both. Thus only half of the triangular depressions in the first layer can be used for placement of 2nd-layer spheres. Triangular holes in the 2nd layer lie directly over the unused holes of the first layer, so that channels pass right through both layers.

The point at the centre of such a channel is surrounded by 6 spheres, arranged as 2 equilateral triangles, one triangle in the 1st and one in the 2nd layer. These 6 spheres form a regular octahedron, a geometrical solid with 6 vertices and 8 equilateral triangular faces.

The space surrounded by the 6 spheres is called an

octahedral hole. An octahedral hole is larger than a tetrahedral hole because 6 spheres cannot be crowded as closely together as 4 spheres.There are 2 ways in which a third layer of spheres can be added to the existing 2 layers in close-packed fashion. In the first way, spheres are place directly over spheres of the first layer. The third and first layers then have spheres in the same locations.

The 4th layer then goes over the second, the 5th over the 3rd, and so on, so a pattern that repeats every other layer is obtained. This is characterized as a 1212... pattern. The unit cell is hexagonal in shape; the 1212... close-packing arrangement is therefore called hexagonal close packing (hcp). Note that the coordination number (the number of nearest neighbors) of any sphere in hcp is 12.

The second way of adding a third layer is to place spheres directly over the octahedral holes between the first two layers. In this case, the third-layer spheres do not lie over spheres in layer 1. This is still a close-packing arrangement because the coordination number of any sphere is 12, just as in hcp. The 4th layer of spheres is then placed directly over the spheres in the first layer.

This type of close packing is designated 123123... to indicate that the pattern repeats after 3 layers. The unit cell for 123... close packing. Although it may not be obvious from the figure, this unit cell is a cube (this is a situation in which a 2-dimensional drawing does not adequately convey what is true). This type of close packing is therefore called cubic close packing (ccp).

Hcp and ccp are the two ways of close packing spheres in 3 dimensions. They have several features in common:

- Spheres occupy 74 per cent of the available space;
- Each sphere has 12 nearest neighbors that touch it (coordination number);
- The number of octahedral holes in the lattice is exactly the same as the number of spheres in the lattice, and there are twice as many tetrahedral holes as spheres.

There are a number of substances that crystallize in one

or the other of these two packing arrangements. Beryllium, magnesium, and zinc metal all form hexagonal close packed lattices. Nickel, copper, gold, and aluminum form ccp lattices. Cubic close packing is particularly common. We will take a closer look at cubic lattices now.

The three Bravais cubic lattices. The simple cubic (sc) lattice is primitive, with points (spheres or atoms) only at the vertices of the cube. Metallic polonium and non-metallic dioxygen (O_2) crystallize with the simple cubic lattice. The coordination number of an atom in the simple cubic lattice is 6. To determine this, pick one atom in the unit cell; find an atom that appears to be closest to the atom you picked; then count the number of other atoms that are at this same distance from the atom you picked.

In counting coordination numbers, you may have to go outside of the single unit cell. In the simple cubic lattice, identical unit cells surround the pictured one on all sides. Some of the nearest neighbors of your picked atom lie in these neighboring unit cells. Finally, the simple cubic unit cell contains only 1 atom, despite the fact that you see 8 atoms drawn in the figure.

We say that the simple cube has 1 atom per unit cell (1 atom/uc). The procedure for counting atoms in unit cells is described in the paragraph below. The body-centered cubic (bcc) lattice is non-primitive, with points at the corners and the centre of the unit cell (the centre of a cube is called the cube body centre). Metallic iron and barium crystallize in the bcc lattice.

The coordination number is 8 for this lattice (count the nearest neighbors of the body-centre atom), and the body-centered cube has 2 atoms per unit cell. Finally, the face-centered cubic (fcc) lattice is also non-primitive with points at each corner and in the centre of each of the 6 faces of the cube. Metallic nickel and copper adopt this structure. The coordination number is 12 for this lattice, which means that it must be close packed! The face-centered cubic lattice and the cubic close packed lattice are identical. It is very difficult to see this in two-dimensional drawings.

To see it, you must imagine looking along a body diagonal of the face-centered cube. A body diagonal runs on a straight line from one corner, through the body centre of the cube, to the diagonally opposite corner. As you look along the body diagonal, you will see the pattern of spheres: a single sphere; a triangle consisting of 6 spheres; a second triangle of 6 spheres, but rotated 60° with respect to the first one; and finally another single sphere immediately below the first one. The cube is admittedly difficult to visualize from the drawings. The face-centered cubic lattice has 4 atoms per unit cell.

We now address the problem of determining the number of atoms per unit cell. This is quite simple once it is realized that atoms are generally shared among several unit cells and therefore must be partitioned accordingly. An atom at a corner (vertex) of a cube is shared among 8 unit cells. To see this, imagine the simple cube in Figure to be a building block, and focus on the upper right front corner of the building block. Imagine that this corner is labelled "a".

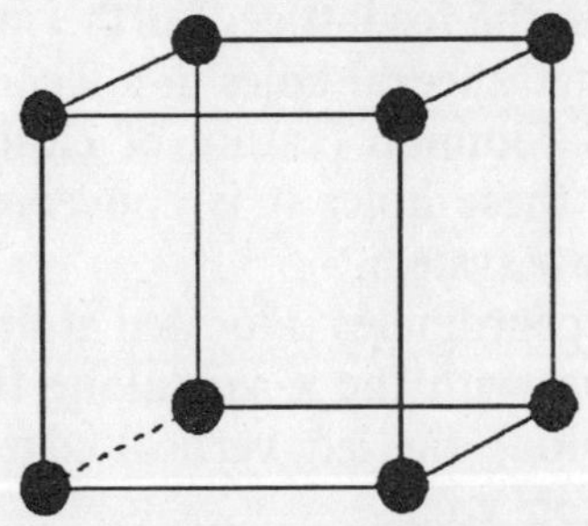

In your mind, place identical blocks on top, to the right, and in front of the block in the figure. Then place additional blocks in the remaining places adjoining these blocks to generate the 8-block. Point "a" is now at the centre of this collection, and is simultaneously a corner for each of the 8 blocks; 8 blocks share a vertex. This point cannot be said to belong entirely to any particular block. Instead, each block gets 1/8 of it. The counting rule for a corner atom, then, is that only 1/8 of it is assigned to a single unit cell.

Similar thinking leads to the conclusion that a face-centre atom is shared between 2 unit cells, so 1/2 of it is assigned to a particular unit cell; and a body-centre atom is not shared at all, so it belongs entirely to the unit cell that it is found in. The atom count for a body-centered unit cell then goes like this:

atoms/uc = 8 corner atoms x 1/8 + 1 body-centre atom x 1

We will have occasion also to partition points located along edges of the cube. Edges are the lines connecting adjacent vertices. Since an edge is common to 4 cubes arranged in a square, 1/4 of any point along an edge is assigned to a particular cube.

Table: Atom Counting Rules for Cubic Unit Cells

Atom location	*Fraction Assigned to uc*	
Corner	1/8	
Edge	1/4	
Face	1/2	
Body-center	1	

If, as we have said, the face-centered cubic lattice is in fact cubic close packed, then there should be tetrahedral and octahedral holes in the fcc lattice. Further, we should find 4 octahedral and 8 tetrahedral holes in a fcc unit cell based on the the last-listed common feature of close-packed lattices above. To locate these holes it is convenient to place our cube in a coordinate system.

The origin of coordinates is located at the lower left front corner of the cube, with the x-axis along the bottom front edge, the y-axis along the left vertical edge, and the z axis along the left bottom edge.

We specify distances along the 3 axes in terms of fractions of the cube edge length. The set of coordinates (1,0,0) means to proceed 1 edge length along x. These are the coordinates of the lower right front vertex of the cube (labelled 1 in the figure). Similarly, (1/2,1/2,1/2) specifies the body centre of the cube, labelled 2 in the figure.

Now examine the face-centered cube. We find the most obvious octahedral hole at the body centre. This point is surrounded by 6 atoms all at the same distance from it. These are the atoms at the face centers of the cube. Similar octahedral holes are located at the cube edge centers.

It is surrounded at equal distance by the atoms at (1/2,1/2,1), (1,0,1), (1,1,1), and (1,1/2,1/2). The remaining 2 atoms are not part of the unit cell shown, but can still be located with coordinate sets (1 1/2,1/2,1) and (1,1/2,-1 1/2). (Translating these coordinates is good practice in the use of the unit cell coordinate system.)

The cube has 12 edges, so there are 12 holes of this type. The counting rules tell us that only 1/4 of each of these edge-centered octahedral holes belongs to the unit cell of interest. The total number of octahedral holes in the unit cell is then 1(body centre) + 12(1/4) = 4. The number of octahedral holes is the same as the number of atoms in the unit cell!

A tetrahedral hole in the fcc unit cell. It is defined by the four shaded atoms, and has coordinates (1/4,1/4,1/4). There are 7 other tetrahedral holes in the unit cell, all with coordinate combinations involving 1/4 and 3/4, and all defined by 1 corner and 3 face-centre atoms. Try to locate them in the figure, and write down their coordinates. Each tetrahedral hole lies entirely within the cube body, so none of them are shared with other unit cells. The number of tetrahedral holes per fcc unit cell is therefore 8, twice the number of atoms in the unit cell.

Octahedral and tetrahedral holes are important in the discussion of ionic compounds.

IONIC COMPOUNDS

The ideas discussed thus far apply to crystals in which the units — spheres or atoms — are all identical. Crystals of pure metals, with close-packed metal atoms, and even crystals of molecular solids (for example, methane, CH_4, in which there is a methane molecule at each lattice point) are examples of real situations in which these ideas apply.

Ionic compounds are different because there are two distinctly different units — the cation and the anion — that must be accommodated in the lattice. The resulting lattices are often called compound lattices because they contain two different components. The cation and the anion differ in charge, of course, and in chemical identity (usually). They also differ in size. Generally speaking, cations are smaller than anions, for easily-understood reasons. Cations contain an excess of protons over electrons. The Coulomb pull of the excess positive charge causes the electron cloud to contract, so that it is smaller than in the corresponding atom (e.g., Na^+ is smaller than Na).

Anions contain an excess of electrons over protons. The repulsion between the electrons exceeds that in the corresponding neutral atom, causing an expansion of the electron cloud (e.g., Cl^- is larger than Cl). Except for a few rare cases, the cation in an ionic compound is smaller than the anion. In forming a lattice, the space requirements are more important for the larger anion, which tends to dictate which lattice is adopted.

The (smaller) cations then fit into the holes, or interstices, of the anion lattice. Many common ionic compounds have lattices that consist of cubic close packed anions, with cations in tetrahedral and/or octahedral holes. Stoichiometries of up to 3 cations per anion can be accommodated in this framework. We now discuss some common ionic lattices.

The Sodium Chloride (Rock Salt) Lattice. It can be described as a cubic close packed (face-centered cubic) anion lattice, with Na^+ ions in the octahedral holes. The coordination number (number of nearest neighbors of opposite charge) of each Na^+ ion is 6, and of each Cl- ion is 6. The cation and anion must have equal coordination numbers because the stoichiometry of the compound is 1:1 (1 cation per anion).

The unit cell contains 4 formula units of NaCl (4 Cl- ions, which define the fcc unit cell; 4 Na^+ ions because there are 4 octahedral holes per unit cell). Many ionic compounds exhibit the NaCl lattice, including most of the group 1 halides, most of the group 2 oxides, and a number of d-metal oxides.

The Zinc Blende Lattice. Zinc blende is the common name for one crystal modification of the common mineral, ZnS. The other form is known as wurtzite. Wurtzite consists of an hcp array of S^{2-} ions with Zn^{2+} ions in one-half of the tetrahedral holes; it will not concern us further here, because we are restricting our attention to cubic lattices. Zinc blende consists of a ccp array of S^{2-}, with Zn^{2+} in half of the tetrahedral holes.

Since there are 8 tetrahedral holes in the sulfide ion lattice, but only 4 S^{2-} ions, only half of the holes are used to accommodate 4 Zn^{2+} ions. This is consistent with the 1:1 stoichiometry of the compound. The zinc ions occupy tetrahedral holes so that they can be as far apart from one another as possible to minimize repulsions between positive ions. If the first Zn^{2+} ion occupies the hole at (1/4,3/4,1/4), the other three will be in holes at (3/4,1/4,1/4), (1/4,1/4,3/4), and (3/4,3/4,3/4). You should locate these holes and verify that the Zn^{2+} ions describe a tetrahedron. The cation and anion coordination numbers are both 4 in zincblende. As for NaCl, there are 4 formula units per unit cell. Other ionic compounds that adopt the zinc blende lattice include BeS, CdS, HgS, and AgI.

The Calcium Fluoride (fluorite) Lattice. This compound has formula CaF_2, and exhibits. The Ca^{2+} ion is virtually the same size as the F- ion, one of those rare situations referred to above, and forms a face-centered cubic lattice. The F- ions fill all of the tetrahedral holes in the cation lattice. Since there are 8 such holes per 4 Ca^{2+} ions, the stoichiometry is nicely accommodated.

The coordination number of the Ca^{2+} ion is 8, and that of the F- ion is 4. (Note that the number of cations per formula unit multiplied by the cation coordination number is equal to the product of the number of anions per formula unit and the anion coordination number.) There are again 4 formula units per unit cell. The fluorite structure is very common for ionic compounds of 1:2 (or 2:1) stoichiometry.

The Cryolite Lattice. Cryolite is the common name for the mineral, Na_3AlF_6. Though rare, this substance is of extreme commercial value, because it is used in the process of extraction of aluminum from bauxite ore. Cryolite consists of Na^+ cations and AlF_6^{3-} anions. It adopts a lattice in which the anions are cubic close packed (fcc) and the cations occupy all of the octahedral and tetrahedral holes. The Diamond Lattice. Diamond is one crystal modification of the element carbon. It is, of course, not ionic, nor even a compound. It is dealt with here because it, too, is based on a fcc unit cell.

The diamond lattice is just like the zinc blende lattice, with carbon atoms in place of both Zn^{2+} and S^{2-} ions. We therefore describe it as a fcc array of carbon atoms with carbon atoms in one-half the tetrahedral holes. The coordination number of each carbon atom in diamond is 4.

The Cesium Chloride Lattice. The cesium chloride lattice. It is a more open lattice than those considered thus far. This lattice consists of a simple cubic array of chloride ions, with Cs^+ ions in cubic holes. A cubic hole is the space at the body centre of a simple cubic unit cell. There is 1 formula unit of CsCl per unit cell, and the coordination number of both cation and anion is 8. Ionic compounds that adopt the CsCl lattice include CsBr, CsI, the halides of thallium(I), and NH_4Cl.

Radius Ratio Rules. Compounds of 1:1 stoichiometry commonly adopt one of three lattices discussed above: the zinc blende lattice, the rock salt lattice, or the cesium chloride lattice. Of interest now is whether or not we can predict, or at least understand, why a particular compound favors one lattice over the others. A number of factors are important in choice of lattice, but one of these dominates in importance: the relative sizes of cation and anion.

Because we view the cation as occupying holes in the anion lattice, we begin by calculating the sizes of the various hole types in relation to the size of the anion involved. The most straightforward calculation is for the octahedral hole. The figure shows one face of the fcc unit cell, with the anions just touching along a face diagonal of the cube. There is an octahedral hole at the centre of each of the 4 cube edges shown; we focus on the right edge.

The size of the octahedral hole will be taken as the distance from the edge centre to, say, the top right anion, along the cube edge. (The distances from the hole to the bottom right anion and to the face-centered anion are the same.) We can express the right edge length as a sum of anion and hole radii as in equation:$a = 2r^- + 2r^+$

Here r^- is the anion radius, r^+ the hole radius. (The + designation is used in anticipation of putting a cation in the hole.) Similarly, we express the length of the face diagonal (fd) in terms of anion radii. This is easy, since the anions touch along this line:

$$fd = 4r^-$$

Finally, we relate the fd and edge length via the Pythagorean Theorem:

$$(fd)^2 = 2a^2$$

Take the square root of both sides of 7-4-3:

$$fd = a(2)^{1/2}$$

Substitute 7-4-4 in 7-4-2 and rearrange for a:

$$a = 2(2)^{1/2}\, r^-$$

Substitute for a in 7-4-1 and solve for r^+:

$$r^+ = ((2)^{1/2} - 1)\, r^- = 0.414\, r^-$$

The octahedral hole is somewhat less than half as large as the anions that create the hole. By similar geometric methods, equations for tetrahedral and cubic holes respectively, can be developed:

$$r^+ = 0.224\, r^- \quad r^+ = 0.732\, r^-$$

You should try to obtain these expressions, making use of right triangles and Pythagorean relationships. (In addition to equation, which relates the face diagonal and edge length of a cube, equation relates the cube body diagonal (bd) to the edge length:

$$bd = (3)^{1/2}$$

A tetrahedral hole is about 1/2 the size of an octahedral hole and about 1/3 the size of a cubic hole. Only cations that are relatively small compared with the anion can be accommodated in tetrahedral holes. As cation size increases relative to anion size, the cation will prefer a larger hole, with a larger coordination number of anions. Equations are the basis for the following radius-ratio rules.

- If the ratio of the cation and anion radii (r^+/r^-) lies in the range between 0.224 and 0.414, the cation prefers to occupy a tetrahedral hole.

- If r^+/r^- lies in the range between 0.414 and 0.732, the cation prefers an octahedral hole.
- If r^+/r^- exceeds 0.732, the cation prefers a cubic hole.

These rules make good sense. If $r^+/r^- < 0.224$, the cation is too small for the tetrahedral hole. The anions will be touching, but will not be touching cations. Since it is the cation-anion attractions that stabilize the lattice, whereas anion-anion or cation-cation repulsions tend to destabilize it, this is not a favorable situation. When $r^+/r^- = 0.224$, the cation just fits the tetrahedral hole. Although anions are still touching, they now touch cations as well. This is a stable arrangement. As r^+/r^- becomes > 0.224, the cation is too big for the tetrahedral hole and starts to push the anions apart.

The cation-anion attractions are maintained, because cations and anions still touch, but now anion-anion repulsions are decreased because the anions no longer touch. This is an even better arrangement than the perfect fit. This situation continues to improve until r^+/r^- becomes equal to 0.414. At this value, the cation prefers to occupy an octahedral hole because it can increase its coordination number without increasing anion-anion repulsion to any significant extent. The number of cation-anion attractive interactions increases from 4 to 6, giving a more stable lattice.

As r^+/r^- increases between 0.414 and 0.732, cations in octahedral holes maintain their contact with 6 anions while pushing the anions further and further apart, decreasing repulsions between them. Throughout this range of r^+/r^-, occupation of an octahedral hole becomes increasingly favorable until the radius ratio becomes 0.732. The cation can now increase its coordination number to 8 anions, again increasing the number of attractive cation-anion interactions. The radius ratio rules are summarized in Table. Radii for common cations and anions are given in Table.

Table: Radius Ratio Rules

Range of r^+/r^-	*Lattice Type*	*Example*
0.224-0.414	ZnS	ZnS
0.414-0.732	NaCl	KCl
> 0.732	CsCl	CsCl

Ion radii and the radius ratio rules can be used to predict the probable choice of lattice for an ionic compound.

Example. The radii of the Na^+ ion and the Br^- ion are 116 and 182 pm, respectively. Which of the three 1:1 lattices will be adopted by NaBr?

Solution. Calculate r^+/r^-:

$$r^+/r^- = 116/182 = 0.64$$

This falls between 0.414 and 0.732. The cation will prefer the octahedral hole. NaBr should therefore have the rock salt structure. It does.

The radius ratio rules provide useful guidelines to the lattice structures of ionic compounds. It is important to realise, however, that they are not hard and fast; the predictions made using the rules are often wrong. For example, in the zincblende form of ZnS, Zn^{2+} ions occupy tetrahedral holes in a fcc lattice of S^{2-} ions. In fact, this compound is the prototype for this type of structure.

However, the radius ratio rules predict that the Zn^{2+} ions should occupy octahedral holes! The radius ratio rules fail frequently because the relative size of the ions is not the only factor that is important in determining structure. Of particular importance is the tendency toward at least some covalent bonding between cation and anion. Covalency involves directional requirements, and may influence the type of hole preferred by the cation.

X-RAY CRYSTALLOGRAPHY

In 1912, von Laue predicted that crystals could serve as diffraction gratings if the wavelength of radiation used were approximately the same as the distance between planes of atoms in the crystals. This spacing was known to be on the order of 1×10^{-10} m, corresponding to wavelengths in the x-ray region of the electromagnetic spectrum. Subsequently, WH Bragg discovered that when X-rays of appropriate wavelength are directed onto the smooth face of a crystal, the X-rays are diffracted and a pattern characteristic of the lattice structure of the crystal is obtained.

This observation is the foundation for the modern

technique of X-ray Crystallography, the most powerful method that we have for determining molecular structure. Like the types of spectroscopy, x-ray crystallography is based on the interaction of light with matter. Unlike spectroscopy, however, the light is not absorbed by the matter in this case. Instead, it is scattered from the atoms of the crystal in a very definite way.

Analysis of the scattering enables the arrangements of atoms within the crystal to be deduced. The analysis of x-ray diffraction patterns is complex, even for materials of simple structure. However, thanks to the efforts of many crystallographers over many years, it is today done quite quickly and routinely.

Once a suitable crystal of a substance is grown, the modern crystallographer is able to mount it in a fully automated x-ray diffractometer, the instrument that is used for the x-ray diffraction study. The diffractometer automatically obtains all of the necessary data for the determination of the lattice structure. The data are then fed to one of a number of computer programs that have been developed and refined for the analysis of such data, and the structure is "solved."

The output of the programme reveals not only the lattice structure, but also the locations of all of the atoms of a molecule of the substance. Structures of substances ranging from simple salts or molecular substances to huge complex proteins and enzymes can be determined using this approach. The structures of two molecules, obtained from x-ray diffraction data. The arrangement of atoms in the relatively simple molecule, phthalocyanine.

Example. Aluminum crystallizes in a cubic lattice. X-ray diffraction reveals that the edge length of the unit cell is 405.0 pm. The density of Al is 2.7 g/cm^3. Which cubic lattice does Al adopt?

Solution. We adopt the following strategy:

unit cell mass $\rightarrow$ Number Al atoms per uc $\rightarrow$ lattice type

We can obtain the unit cell mass from the density by calculating unit cell volume from the cube edge length.

Volume of uc = $a^3 = (4.05\times10^{-8}\ cm)^3 = 6.643\times10^{-23}\ cm^3$/uc

Mass of unit cell = r x Volume uc = $2.7\ g/cm^3\times6.643\times10^{-23}\ cm^3/uc = 1.794\times10^{-22}$ g

Moles Al per uc = mass/molar mass = 6.647×10^{-24} moles AlAtoms Al per uc = 6.647×10^{-24} moles$\times6.022\times10^{23}$ atoms/mol= 4.0

The unit cell is face-centered cubic. This example shows the connection between macroscopic and microscopic properties. Note that in a reversal of the above procedure, knowledge of the type of cubic lattice could be used to experimentally measure Avogadro's number, N_o.

Example. In early nuclear reactors, fuel rods were made of uranium metal, with a cubic close packed lattice. These suffered from the problem that the fuel rods frequently burst out of their protective sheaths, which could obviously cause major safety problems. Why did this occur, and what could be done to prevent it?

Solution. Nuclear fission (the spontaneous breakup of a radioactive nucleus) produces 2 nuclei for each one that decays. One possible decay route is shown below:

$$^{235}U \rightarrow {}^{90}Sr + {}^{143}Xe + 2\ {}^{1}n$$

In the ccp U lattice, the holes are too small to accommodate these extra nuclei. The resulting expansion of the lattice caused bursting of the protective sheaths. The solution of the problem was to make the fuel rods from UO_2 (yellow cake), which consists of ccp U atoms with O atoms in the tetrahedral sites. The larger octahedral holes in this lattice were able to accommodate the fission fragments without expansion.

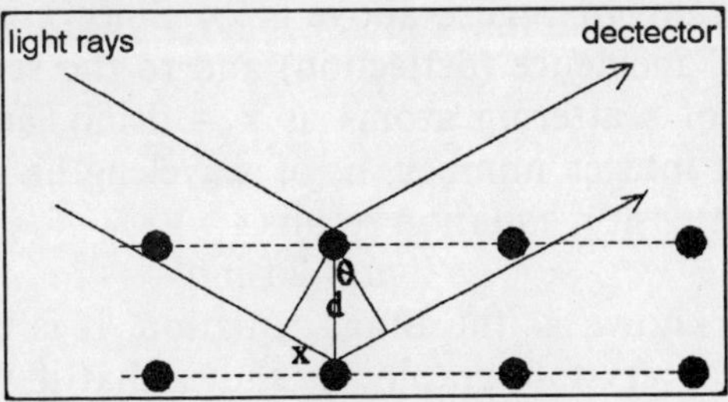

The Bragg Equation. Figure shows two planes of atoms in a crystal. A source of X-rays at the left emits radiation, two beams of which are shown impinging on the planes of atoms and scattering off to the right, where they are "seen" at a detector.

The angles made by the incident beam and the deflected beam with the planes of atoms are the same. A diffraction pattern is a series of alternating light and dark lines (sometimes spots). The white spots result from the arrival at the detector of reflected beams that are all in-phase. That is, their waves are matched up peak for peak and trough for trough. Waves that are in phase add together to give a wave with larger peaks and larger troughs — i.e., more intensity. This is called constructive interference.

The white spots in the diffraction pattern correspond to deflection angles that give in-phase waves. Similarly, the dark areas correspond to angles that give out-of-phase waves. Here the peak of one wave arrives at the detector at the same time as a trough of a second wave arrives. Such waves cancel each other, and very little or no signal is seen at the detector. This is called destructive interference.

The criterion that must be met in order that two waves arrive at the detector in phase is that the difference in the distances travelled by them must correspond to an integer number of wavelengths. This will give the maximum intensity at the detector. If the difference in distance travelled corresponds to a half-integral number of wavelengths, cancellation will occur, and there will be no intensity registered at the detector.

The difference in the distances travelled by the two deflected beams in Figure above is 2x. Since x is related to the angle of incidence (deflection) and to the separation of the planes of scattering atoms as x = dsinq, and since 2x must be an integer number, n, of wavelengths in order to maximize intensity, equation results.

$$n = 2d\sin q$$

This is known as the Bragg equation. It is the basis of modern x-ray crystallography. It shows that if we measure the angles at which constructive interference occurs, we can calculate d, the spacing between atom planes in the crystal.

Example. A set of planes in a crystal is separated by 250 pm (2.50×10^{-10} m). What is the maximum x-ray wavelength that could be used to measure this distance?

Solution. The largest scattering angle possible is 90°. Since sin 90° = 1, n = 2d = 5.00×10^{-10} m. q is largest for n = 1, giving = 5.00×10^{-10} m. A wavelength longer than this cannot fit even once into the distance, 2d, so cannot be used to measure the distance. X-rays with less than 5.00×10^{-10} m must be used.

Lattice Imperfections

Perfect crystals, in which every atom or molecule is in place, are an idealization, like the ideal gas. Real crystals contain fairly large numbers of imperfections, called defects. Three major types of defect are called point defects, line defects, and plane defects. Common point defects in simple lattices (in which all lattice points are identical) are vacancies (in which an atom is missing); atoms out of place, occupying what is normally a hole in the lattice; substitutional impurities, in which foreign atoms occupy occasional sites; and interstitial impurities, in which foreign atoms occupy holes in the lattice.

Ionic (compound) lattices exhibit Schottky defects, in which a cation or anion is missing; and Frenkel defects, in which an ion leaves its site and enters a lattice hole. Line defects result from missing rows of atoms. Particularly common are edge defects, in which an extra partial plane of atoms is inserted near an edge of the crystal.

Example. Bending a paper clip. Take an ordinary paper clip and try to straighten it out. Are you able to get it perfectly straight in the regions where it was bent? Why not? Now try to rebend the paper clip to its original shape. Are you able to do it? Why not?

Solution. When the clip was manufactured, it was annealed following bending. Annealing is the process of heating the clip to soften it enough that atoms can diffuse into the gaps left in the bending process. In other words, annealing produces edge defects. When you attempt to straighten the clip, the edge defects prevent the alignment of atomic planes, so complete straightening is impossible. However, straightening the clip introduces new edge defects on what was originally the inside curve of the bend, so that when you attempt to rebend the clip, the new defects interfere.

Plane defects in crystals occur at the interface between two differently-oriented crystalline regions. As might be expected, such defects greatly diminish the resistance of the crystal to fracture. Crystal defects are important for a number of reasons. A problem with defects, especially of the line and plane type, is that they diminish the structural strength of the solid. Annealing is used to offset this.

Rusting of iron often occurs at defect sites on the surface, which may result when the metal is worked or penetrated by a rivet or bolt. However, defects can be advantageous as well. Defects are sites of high energy (i.e., relative instability).

Reactions that take place at the surfaces of crystals often take place at these sites. This is involved in heterogeneous catalysis by solid materials, which is used on a huge industrial scale in our economy. Electronic components for modern computers require the reliable production of silicon that has a purity of 99.999999 per cent. Silicon crystals of such purity contain very few defects, which would diminish their functional effectiveness.

Silicon of this purity is produced by a process called zone refining. In this method, a bar of silicon of high purity is passed very slowly past a very narrow heating element, which melts a narrow region of the silicon bar. Impurities in the silicon dissolve in the melted part, and follow along with the melted region as the bar moves past the heating element. Eventually the fused region, containing all of the impurities, reaches the end of the bar, where it is solidified and cut off. The remaining silicon has the purity required in the manufacture of electronic components.

PROPERTIES OF LIQUIDS

We take up the liquid phase last because it is the most

complex and least understood of the phases of matter. Liquids have properties intermediate between those of the solid and gas phases:

- A sample of liquid has a fixed volume (like a solid) and a molar volume only slightly greater than that of the solid. However,
- A liquid conforms to the shape of its container, like a gas, and
- A liquid flows readily, like a gas.
- Liquids are relatively incompressible. The relative compressibilities of the three phases are in the order g >> l > s. The incompressibility of liquids finds practical use in hydraulic machinery, such as automobile brakes.
- Diffusion is slower in liquids than in gases, but faster than in solids. The relative rates of diffusion in the three phases are in the order g >> l > s.

  These properties are readily observed visually. Several other properties of liquids are more subtle:
- Liquids give x-ray diffraction patterns that are diffuse, rather than sharp like those of solids. The pattern implies some order in the molecular arrangement, but less than in the solid.
- Liquids exhibit surface tension.
- Liquids exhibit viscosity.
- Liquids in closed containers evaporate to an extent. The resulting vapour exerts a pressure called the vapour pressure, P_{vap}, of the liquid. The magnitude of P_{vap} for a particular liquid depends only on temperature, not on the container size or the amount of liquid present. Liquid and vapour exist in equilibrium.

We begin with a molecular model for the liquid phase, based on its observable properties.

MODEL OF THE LIQUID PHASE

The small molar volume and incompressibility of liquids suggest that the molecules are close together. However, flow and the lack of a fixed shape indicate molecular mobility, despite the closeness. A feature of the model that is not evident from the figure is that the molecules are in constant motion, as in the gas and solid, colliding frequently with neighbors. As for the gas and the solid phases, the kinetic energies of the particles of the liquid are described by the Maxwell-Boltzmann distribution, and the average KE per particle is 3/2 kT. Each molecule is fairly closely surrounded by other molecules, but there are definite gaps (holes) in the structure. Molecules use these gaps to slip and slide easily past one another, which manifests macroscopically in the ability to flow.

There are regions in the liquid that are quite ordered, similar to the solid, but the regions are constantly shifting position as molecules move to close gaps and open new ones. The gaps prevent the order from extending over long distances.

Thus liquids have short-range order, in contrast to the long-range order of crystalline solids. The gaps allow diffusion of a molecule through the liquid, but frequent collisions with neighboring molecules makes diffusion slow.

Contrast this with the rapid diffusion in the gas phase, where a molecule travels a long distance between collisions; and the extremely slow diffusion of the solid phase, where a molecule is locked into its lattice location. Diffusion rates in solid, liquid, and gas are best understood in terms of the mean free path, the average distance travelled by a molecule between collisions. The mean free path in the solid is virtually zero. In the liquid, it is a fraction of the molecular diameter. But in the gas, the mean free path may be 10-100 molecular diameters, depending on gas pressure.

INTERESTING PHYSICAL PROPERTIES DIFFUSE X-RAY DIFFRACTION PATTERNS

Indicates regions of short-range order in the liquid. Within these regions, molecules are almost close-packed. However, the gaps allow fluctuation in the distance of separation of the molecules. The nearly close-packed regions of the liquid diffract x-rays according to the Bragg equation, n = 2dsin*theta*. Fluctuations in d, a consequence of the gaps, cause fluctuations in *theta*, giving x-ray patterns that are diffuse, or smeared out.

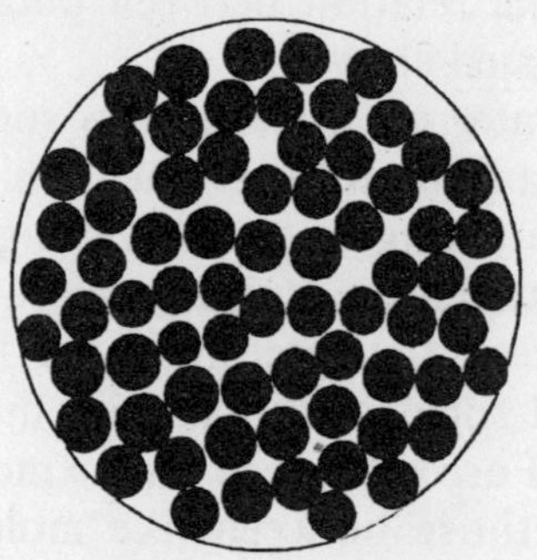

Surface Tension

This is a consequence of intermolecular forces. Consider a molecule at the surface of a liquid. In contrast to a molecule in the bulk, which is attracted equally from all directions, the surface molecule suffers an imbalance of forces because there are no molecules above it. The net force on it is directed toward the interior of the liquid. This pulling in of the liquid surface is called surface tension, symbolized z. The surface molecule has higher potential energy than a bulk molecule because of the lesser force on it.

To minimize energy, the liquid adopts a shape that minimizes the number of surface molecules. The shape with the smallest surface area-to-volume ratio is the sphere. Thus surface tension explains the spherical shape of raindrops and water droplets on a freshly waxed car: the spherical shape minimizes potential energy.

It is a safe prediction that manufacture of ball bearings will someday be carried out in space, where gravitational forces will not distort the sphericity that surface tension imposes on a drop of molten metal.

As temperature increases, molecular kinetic energy can overcome the attraction of intermolecular forces, and surface tension decreases according to equation.

$$z = a - bT$$

The constants a and b are characteristic of a particular liquid. For ethanol, $a = 24.05 \times 10^{-3}$ J/m^2 and $b = 0.0832 \times 10^{-3}$ J/m^2-°C. The units of surface tension are energy per area, consistent with the relation between potential energy and surface area discussed above.

A drop of water on a clean glass surface spreads and flattens, forming a film on the surface. This phenomenon of spreading and flattening is called wetting. Wetting occurs because forces between the surface molecules and molecules of the liquid exceed the intermolecular forces within the liquid. The liquid spreads to maximize the area of contact with the surface. Forces between unlike molecules are called adhesive forces; those between like molecules are called cohesive. Wetting occurs because adhesive forces exceed cohesive forces.

If the reverse is true, a liquid will not wet a surface, but instead will sit on the surface in spherical droplets to minimize the area of contact with the surface. One visible consequence of wetting is the formation of a meniscus at the top of a column of liquid in a tube, such as a buret or pipet. The menisci formed by water and mercury.

The adhesive forces between the glass surface and water molecules are stronger than the cohesive forces between water molecules. Consequently, water achieves a maximum surface area of contact with the glass by bending up at the edges and sagging in the centre. The reverse is true for mercury, which minimizes the surface area of contact with the glass and forms a spherical surface shape to maximize cohesive forces.

The phenomenon of capillary action results from competition between surface tension and wetting. When one end of a capillary tube (a glass tube with inner diameter of approximately 1mm) is submerged in a liquid, a difference in height develops between the levels of liquid in the capillary and in the bulk liquid. A liquid that wets glass rises in the tube; a liquid that does not wet glass falls, called capillary action, for water and mercury. As water wets the inside wall of the capillary, the surface area of the water is increased.

However, surface tension tends to minimize the number of molecules at the surface, and causes the water to move up the tube, against the force of gravity, to offset the surface area increase resulting from wetting. The result is a capillary rise. For mercury, adhesive forces are weak, so the level in the capillary falls to minimize contact with the glass surface, and the surface of mercury in the capillary is pulled into a sphere by the strong cohesive forces. The result is capillary fall. Capillary action is important in the operation of blotters and sponges and is the mechanism by which water penetrates soils.

Viscosity.

The viscosity of a liquid is a measure of its resistance to flow. A mobile liquid is one that flows readily (e.g., water). A liquid that flows with difficulty (e.g., molasses) is said to be viscous. Liquids composed of small, roughly spherical molecules are mobile, because the molecules can roll over one another smoothly, like ball bearings. Viscous liquids are composed of long, string-like molecules that easily become tangled during flow.

For example, the molecules of high viscosity motor oils consist of long chains of carbon atoms that intertwine and impede flow of the liquid. An extreme example of molecular tangling occurs in silly putty, a polymer that appears to be a solid but is in reality a liquid in which molecular tangling is so severe that hours are required for the putty to spontaneously change shape to that of the container. Like surface tension, viscosity decreases (flow becomes easier) as temperature increases due to more rapid thermal motion of the molecules.

Vapour Pressure.

Vapour pressure is the most fascinating physical property of liquids. It provides remarkable insight into the laws governing the behaviour of physical systems. We begin with a purely experimental approach to the phenomenon. The box is attached to a closed-end manometer and is connected via a stopcock to a vacuum pump.

A thermometer is inserted in the box and there is a septum-capped port (opening) that can be penetrated with a syringe needle. The stopcock to the vacuum pump is opened and the air is pumped out until the two arms of the manometer are at the same level. The stopcock is then closed. This is the initial situation, state 1, shown at the top left of the figure. Several experiments will now be performed in sequence.

Vapour pressure can be summarized as follows:

- A liquid in a closed container evaporates to some extent. The resulting vapour exerts a characteristic vapour pressure, P_{vap}.
- The value of P_{vap} is independent of the container size and the amount of liquid, provided that the amount exceeds a critical minimum.
- P_{vap} increases steeply with T.
- P_{vap} is a function only of the chemical identity of the liquid and the temperature.
- Liquids have a natural tendency to evaporate to some extent to form vapour.

The thought experiments, which can be verified readily in the laboratory, provide much descriptive information about the vapour pressure phenomenon. But the questions of why liquids tend to evaporate, and why they establish characteristic vapour pressures.

THE DISSOLVING POWER OF LIQUIDS

Liquids are invaluable to chemists as solvents. They are able to accommodate molecules of solute readily within the gaps in their structures to produce solutions that are themselves liquids. Such solutions are readily poured or otherwise transferred from one container to another; are conveniently measured out quantitatively using volumetric apparatus; and provide a uniform medium in which a chemical reaction can occur.

The unique ability of liquids to act as solvents is due to their dissolving power — the molecular motion of the liquid dislodges ions or molecules from the surface of crystals of a solid, then accommodates the separated ions or molecules in gaps. In this section we briefly discuss the dynamics of the dissolution process at the molecular level.

A microscopic view of two adjoining faces of a crystal of NaCl. A Na^+ ion on a face of the crystal differs from those in the interior because it experiences forces of attraction from only 5 nearest-neighbour Cl^- ions, rather than 6. Facial Cl^- ions are in a similar situation. Such ions are therefore in shallower potential wells than ions in the interior of the crystal and should be more easily dislodged.

Edge ions, such as the Na^+ ion labelled "a" in the figure, are attracted by only 4 nearest neighbors, and should be even more readily dislodged than face ions. Water molecules from the solvent continually bump against the faces and edges of the solid crystal. As they approach the crystal, their dipoles become oriented due to interactions with the face ion that they approach. Water molecules approaching cations have the O atom oriented toward the crystal face; those approaching anions put a H atom forward.

When a rapidly moving water molecule strikes a face or edge ion, sufficient kinetic energy can be transferred to dislodge the ion, which moves away from the crystal into a gap in the solvent structure (e.g., Na^+ ion "b" in the figure).

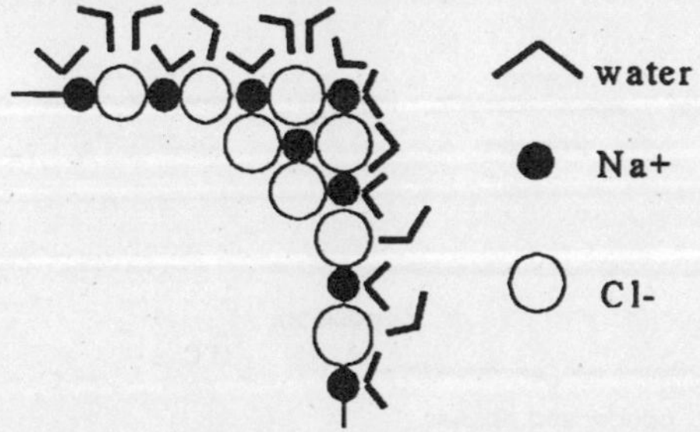

It becomes immediately surrounded by a cage of water molecules, with oxygen atoms nearest the Na^+ ion to maximize ion-dipole forces. The solvent cage insulates it from the crystal and makes return to the crystal face unlikely. Since edge ions are more readily dislodged than facial ions, and because they can be approached by solvent from several directions, dissolution occurs more rapidly at crystal edges than faces. The initially sharp edges of a dissolving crystal become rounded as dissolution proceeds.

PHASE TRANSFORMATIONS

A phase transformation is a conversion of a pure substance from one phase to another. There are 6 types, each involving two of the three phases:

- Melting (fusion) — conversion of solid to liquid.
- Freezing — conversion of liquid to solid; the reverse of melting.
- Vaporization — conversion of liquid to vapour (gas). Its reverse is
- Condensation — conversion of vapour to liquid.
- Sublimation — conversion of solid to vapour. Its reverse is
- Deposition — conversion of vapour to solid.

The phase changes are illustrated in Figure. Since the solid and liquid phases have small molar volumes due to close packing of molecules, they are referred to as condensed phases.

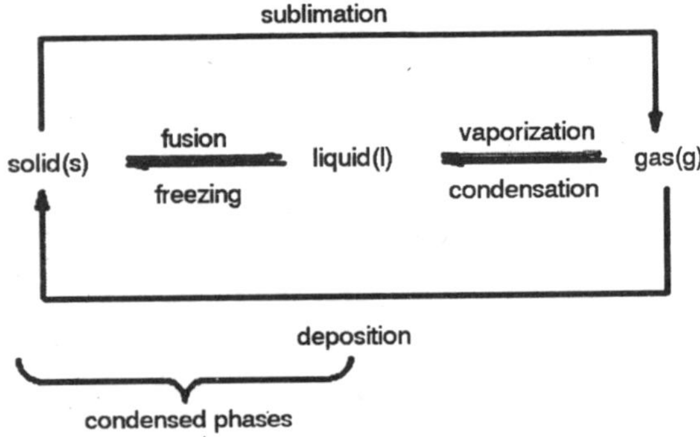

Fig. Phase Transformations

The two variables that determine the phase in which a substance exists are the temperature and the pressure to which the substance is subjected. Phase transformations may therefore be accomplished in two limiting ways: by changing the temperature of the substance while keeping it under constant pressure; or by changing the pressure on the substance while keeping it at constant temperature. The first approach is the more familiar of the two. We make three assertions about temperature-induced phase changes:

- At a constant pressure, a phase transformation occurs at a constant temperature that is characteristic of the substance.
- To convert a given amount of substance from solid to liquid to gas requires addition of energy. This is usually accomplished by adding heat. The energies of the phases are therefore in the following order: $E_s < E_l < E_g$
- Conversion from solid to liquid to gas decreases the order (degree of organization) of the substance.

To clarify these assertions, we take a thought experimental approach to the phase changes of a substance, much as we did for vapour pressure. The results of these thought experiments can be readily verified in the laboratory. We start with a quantity of solid water (ice) in a syringe, state 1. The initial temperature of the ice and the inside of the syringe is -10°C. We plan to slowly add energy to the ice by controlled heating, which can be done by submerging the syringe in a large antifreeze bath with a hot plate under it. As we slowly add heat, we monitor the temperature of the water in the syringe with a thermometer inserted through an air-tight seal in the syringe. We begin at state 1.

Process: Turn on the hot plate and the bath stirrer. As the ice warms slowly from -10 to 0°C, nothing seems to happen. Temperature changes slowly and steadily. When the ice reaches 0°C, it begins to melt, and we see some liquid water form in the syringe.

Process: We continue to gently heat the bath. The ice gradually melts, but the temperature of the ice-water mixture remains constant at 0°C, despite the addition of heat. The temperature of the bath rises above 0°C, but as long as both ice and liquid water are in the syringe, T in the syringe remains at 0°. State 3 is reached when the last ice crystal melts. All water in the syringe is liquid, at T = 0°C.

Process: Once the ice has melted, the temperature of the water in the syringe begins to rise. As we slowly warm the bath, the water temperature slowly rises. Nothing happens in the syringe other than a slight increase in the volume of the liquid until T = 100°C. Then water begins to vaporize. This is the situation at state 4.

Process: We continue to slowly add heat to the bath and observe that temperature in the syringe remains constant at 100°C as long as both liquid and vapour are present, even though the bath temperature rises above 100°C. We also observe that as vapour forms, it pushes the plunger out. After all liquid has vaporized, at state 5, the temperature of the water in the syringe once more starts to rise.

Process: Continue heating the bath to T = 110°C. The temperature of the vapour in the syringe rises from 100° at state 5 to 110° at state 6, while its volume increases according to the ideal gas law. The pressure of the water vapour is maintained constant at 1 atm by movement of the syringe plunger.

These experiments verify the assertions at the beginning of the section. First, phase changes occur at constant T. This is shown by processes 2 and 4. The fixed, characteristic temperature at which a substance melts is called the melting point (s to l) or freezing point (l to s), and is symbolized T_f. The normal melting point is the temperature of melting under a pressure of 1.00 atm. For water, this is 0°C.

The characteristic temperature at which a substance boils under 1.0 atm pressure is called the normal boiling point. For water, this is 100°C. Second, process 2 shows that an input of energy is required to convert a solid to liquid. The same is true for conversion of liquid to gas (process 4). For each of these processes, the change in energy, DE ($E_{final}-E_{initial}$) is positive. Processes 2 and 4 may be represented as follows:

$$H_2O(s),\ 0°C \rightarrow H_2O(l),\ 0°C\ [DE > 0]$$

$$H_2O(l),\ 100°C \rightarrow H_2O(g),\ 100°C\ [DE > 0]$$

The process in equation occurs at constant T. Therefore the change in kinetic energy of the molecules is 0 (recall that an average molecule has KE of 3kT/2 at temperature T, regardless of phase). DE is therefore a change in potential energy, DPE. Thus the PE of liquid is higher than that of solid water at 0°C. Similarly, the PE of vapour is higher than that of liquid water at 100°C. Combining these conclusions gives equation:

$$PE_g > PE_l > PE_s \text{ at any } T$$

DPE > 0 for sublimation, fusion, and vaporization; DPE < 0 for the reverse processes. Third, the amount of disorder at the molecular level increases in our sequence of processes (we phrase this in terms of disorder rather than order). With addition of thermal energy (heat), the molecules convert from the long-range organized arrangement of the solid to the completely chaotic and random molecular distribution of the gas. Even warming the gas from 100 to 110°C decreases the order, because the volume available for the molecules to bounce around in increases. This conclusion is summarized in equation.

$$\text{disorder s} < \text{disorder l} < \text{disorder g}$$

Phases and the Potential Well, Revisited

The origin of the potential well in terms of electromagnetic forces within or between the fundamental units of substances. We concluded that forces are strongest and potential energy is minimized when fundamental particles form an orderly close-packed arrangement, and we recognized this as the situation in the solid phase. Heating the solid phase increases the kinetic energies of the particles of the solid phase until they are able to partially overcome the close-packed forces and separate somewhat. At this point, solid melts to liquid. Separation has increased the potential energy of the particles.

Further heating eventually gives the particles sufficient kinetic energy that they may escape the forces in the liquid into the space above the liquid; i.e., vaporization occurs. In the vapour phase, molecules are so widely spaced that forces between them are negligible and potential energy has increased to zero. We thus see that phase conversion from solid to gas can be viewed in terms of particles climbing out of the potential well. Similarly, in the reverse conversion, particles fall into the potential well. The decrease in potential energy of the molecules is released as heat.

Chapter 8

Physical and Chemical Processes

THE SECOND LAW OF THERMODYNAMICS

We introduced the concept of disorder by discussing the free expansion of an ideal gas. Common experience tells us that this process is spontaneous—it occurs without outside assistance. The change in potential energy for this process is zero, so its spontaneity is not due to the tendency to minimize PE. Instead, the process is driven by the tendency to increase disorder. Our earlier discussion of disorder was vague in two respects.

First, we did not distinguish between the disorder of the system and that of the surroundings. Second, disorder was treated only qualitatively. We will put the disorder concept on a much firmer basis by making it quantitative.

THE SECOND LAW OF THERMODYNAMICS

The free expansion of an ideal gas, expansion of the gas (the system) increases its disorder because each gas molecule has a larger volume to explore. Disorder is greater because knowledge of the location of a gas molecule is less. Clearly, then, free expansion leads to an increase in disorder—in entropy—of the system: $DS_{sys} > 0$. What happens in the surroundings when the gas expands? Other than the small effort expended in opening the connecting valve, absolutely nothing happens.

No boundary of the system moves, so the surroundings does not experience work. There is no heat flow to or from the surroundings because the gas is ideal and thus does not change internal energy on expansion. Absolutely nothing is experienced by an observer in the surroundings. It therefore follows that the disorder in the surroundings is unchanged by the free expansion of the gas: $DS_{surr} = 0$. The system and surroundings together constitute the universe. Combining our two results gives equation for the free expansion of an ideal gas:

$$DS_{sys} + DS_{surr} = DS_{univ} > 0$$

Stated in words, equation says that in any spontaneous process, the entropy of the universe increases. Although the equation was developed by consideration of a single simple process, it has proved to have far-reaching and profound generality: it is true for any spontaneous process. Thus for the burning of natural gas; the rusting of iron; the respiration of living organisms; and the light-promoted process of photosynthesis, the disorder of the universe increases. The simple and powerful statement in Equation is called the Second Law of Thermodynamics.

The Second Law is in many respects more subtle and difficult to understand than is the First Law. Yet it is the Second Law that allows prediction of spontaneity, so to gain some understanding of it is crucially important.

To facilitate understanding, we will make liberal use of examples the important ideas; and we will set down a number of simple guidelines for predicting the sign of the entropy changes for system and surroundings in physical and chemical processes. These guidelines are not hard and fast rules—instead, they give the correct result most, but not all, of the time. However, their simplicity makes them worth remembering.

The search for a quantitative measure of entropy was a long and difficult one. The analysis of steam engines (the original genesis of thermodynamics) was the source of one approach, formulated by Rudolf Clausius in the 1880s. The Clausius formulation expresses entropy in terms of heat and temperature. The second approach, that of Ludwig Boltzmann, was based on considerations of probability and statistics. Although the two quantitative formulations of entropy appear different, it has been shown that they are equivalent in any application to which they may both be applied. We will briefly consider these two approaches to entropy, beginning with the Clausius formulation.

ENTROPY AND HEAT

That there is a connection between disorder and heat is somewhat expected. First, heat, the random thermal motion of molecules, is clearly the most disordered and non-useful form of energy. Second, much of our experience indicates that disorder is produced when heat is added to systems. The solid form of a substance (water, sodium chloride, iron) melts, and the liquid form boils, when heat is added. Gases expand when heated.

Atoms form a plasma of nuclei and electrons when heated to the temperatures of stars. These qualitative realizations help us to appreciate the Clausius definition of entropy in equation.

$$DS_{sys} = q_{max}/T$$

Stated in words, the equation says that the change in entropy in a system when a process is carried out may be calculated by dividing the amount of heat added to the system by the temperature of the system. Strictly speaking, equation is valid only when the temperature of the system remains constant. Most of the processes that we will consider occur at constant temperature, however, and equation applies.

The subscript "max" on the heat term represents a restriction on the equation. It indicates that when calculating the entropy change for a change in the system from a particular initial to a particular final state, we must use the heat that would be added to the system along the maximum heat path. This is known as the reversible path.

Our intention here is merely to gain appreciation for the connection of entropy to heat. Consequently, we will not discuss this idea of maximum heat further here. We now proceed to examine the "sense" of equation. First, the Clausius formulation of entropy recognizes the clear connection between disorder and heat. The more heat we add, the more disorder we produce, at a given temperature. Second, the equation says that the disorder produced by addition of a given amount of heat is less when the system is at high temperature than when it is at low temperature. This is also reasonable. In a system at low T (say, near absolute zero) the kinetic energy of random molecular motion is small. At high T (say 1000 K), however, the thermal energy is substantial.

Addition of a specified amount of heat to such a system produces a much larger fractional change in disorder at low T than at high T. Equation recognizes that if a system is already highly disordered, a small amount of heat will not make much difference.

For example, when your room has not been cleaned for a month (or longer), the addition of one more dirty sock to the mess will hardly be noticeable.

In contrast, if your room has just been cleaned (lowered in entropy), that dirty sock on the floor will be very noticeable. Third, the equation tells us that the units of disorder are those of energy divided by temperature, or J/K. Finally, entropy is represented in equation by an upper case symbol, which we reserve for state functions. Entropy is a state function.

This means that the change in entropy of a system is independent of the path taken between initial and final states. Two simple and useful guidelines can be distilled from equation. Both guidelines refer to systems consisting of a single pure substance.

Guideline: If heat is added to a pure substance, the entropy of the substance must increase.

Guideline: If heat is removed from a pure substance, the entropy of the substance must decrease.

Example: In which member of each pair is entropy (disorder) higher?

Fe(s) at 0 °C; Fe(s) at 25 °C

$H_2O(l)$ at 298 K; $H_2O(g)$ at 398 K

Ar(g) at 300 K; Ar(g) at 150 K

Solution: To convert Fe(s) at 0 °C to Fe(s) at 25 °C requires the addition of heat. As a result of the addition of heat, the average kinetic energy of the iron atoms increases, with a corresponding increase in disorder in the system. Fe(s) has higher entropy at 25 ° than at 0 °C. To convert $H_2O(l)$ at 298 K to $H_2O(g)$ at 398 K requires the addition of heat. First, liquid water must be heated from 298 K to the boiling point, 373 K.

This increases the thermal motions, and the disorder, of the molecules of the liquid. Second, liquid water must be boiled at 373 K. This frees water molecules from the potential well of the liquid and allows them to move independently of one another; disorder increases substantially. Finally, water vapour at 373 K must be heated to 398 K. Again, average molecular kinetic energy (and disorder) increases with the addition of heat. $H_2O(g)$ at 398 K has substantially higher entropy than $H_2O(l)$ at 298 K.

To convert Ar(g) at 150 K to Ar(g) at 300 K requires the addition of heat. This increases the random thermal molecular motion, and the disorder of the system. Ar(g) has higher entropy at 300 K than at 150 K.

Example: When 15.0 g of liquid ethanol is vaporized at its normal boiling point of 78 °C, 12810 J of heat is absorbed by the system from the surroundings. What is the change in entropy of the system in this process? Recall that the normal boiling point is the temperature at which the vapour pressure of the liquid is 1 bar.

Solution: Because heat is added to a pure substance in this process, we can say with confidence that the entropy of the system (substance) increases. Thus $DS_{sys} > 0$. In this case, we can go further by actually applying equation to calculate the change in entropy that occurs when ethanol is vaporized, because phase changes carried out at the normal phase change temperature satisfy the "maximum heat path" restriction.

$$DS_{sys} = q_{phase\ change}/T_{phase\ change}$$

Further, vaporization at the normal boiling point implies a constant pressure of 1 atm. In constant pressure processes, q = DH. Thus

$DS_{sys} = DH_{phase\ change}/T_{phase\ change} = 12810\ J/351\ K =$ 36.5 J/K

Example leads to a third useful entropy guideline:

Guideline 3: DS_{sys} for a phase change at the normal

phase change temperature is given by

$$DH_{phase\ change}/T_{phase\ change}$$

ENTROPY AND PROBABILITY: THE BOLTZMANN INTERPRETATION OF ENTROPY

The Clausius interpretation of entropy is a pragmatic one, based on engine analyisis. Ludwig Boltzmann took a different aproach, based on the idea that there is a parallel between the amount of disorder in a system in a particular state and the number of ways that the system may be arranged in that state. In a very general way, this can be stated as

$$S = f(W)$$

where W = the number of arrangements possible for a particular state of a system, and f is a to-be-determined function. We suppose at the outset that a state with only one possible arrangement has no disorder—that is, S = 0—and that S should increase smoothly as W increases. Thus our first constraint on the form of the function, f, is that f(1) = 0. To arrive at a second constraint on f, consider a simple example. Shown is a row of 10 buckets placed side-by-side.

A ping pong ball is tossed at the row of buckets and lands in one of them (we assume for simplicity that there are no misses—the ball always goes in one of the buckets). The state of the resulting system is described by saying that there is one ball in the set of 10 buckets.

Because the ball can be in any of the buckets, the number of arrangements for the state, W, is 10. There is thus some disorder associated with this state. An observer asked to view the row of buckets containing one ball would be uncertain as to where the ball is. His uncertainty is a measure of the disorder.

We now double the amount of material in the system by tossing a second ball, which also lands in one of the buckets. The state of the system is described by saying that there are 2 balls in the set of 10 buckets. There are 10 possible locations for the first ball. For each of these possibilities, there are 10 possible locations for the second ball. The total number of arrangements for the state is $10\times10 = 10^2$. By similar reasoning, if n balls in succession were tossed, the number of arrangements would be 10^n. In general, for a state consisting of N balls placed in L locations, the number of arrangements is given by equation.

$$W = L^N$$

The entropy of the state is $S = f(L^N)$. We now impose the second constraint on the form of the function f: S must double when the amount of material in the system (N) doubles. In equation form,

$$f(L^{2N}) = 2f(L^N)$$

The function that satisfies both imposed constraints ($f(1) = 0$ and $f(L^{2N}) = 2f(L^N)$) is the logarithm. Thus

$$S \text{ proportional to } \ln W = C \ln L^N = NC \ln L$$

where C is a proportionality constant. Equation is the Boltzmann interpretation of entropy. According to equation, the entropy of a system is directly proportional to the number of particles (such as ping pong balls, or molecules), N, in the system and to the natural logarithm of the number of locations each particle can have.

Thus the amount of disorder goes up directly with the amount of matter in the system, in agreement with the second constraint. When each particle has only one allowed location, L is 1 and S is 0. Further, we see that as the number of allowed locations for a particle (L) goes up, S increases more and more slowly. This is seen below:

L	*S*	*Increment in S*
1	0	
2	.693 NC	.693 NC
3	1.10 NC	.41 NC
4	1.39 NC	.29 NC
5	1.61 NC	.22 NC
.		
.		
99	4.60 NC	
100	4.61 NC	0.01 NC

When the system has low entropy initially (L is small, say 1), the addition of one more allowable location has a dramatic effect on S. Thus increasing L from 1 to 2 increases S by 0.693 NC. However, when the system has high entropy initially (L is large, say 99), increasing L by 1 increases S by only 0.01 NC, a negligible effect.

Boltzmann expressed equation in somewhat different form, $S = k \ln W$

The constant, k, is called Boltzmann's constant. It has the value 1.38×10^{-23} J/K. A plot of S versus W, the number of arrangements of N particles in L locations. The slope of the plot is steep when L is small, but approaches zero when L becomes large, showing that DS becomes smaller for a given value of DW as S (and W) increases. Thus when W increases from 1 to 200, S increases by about 7×10^{-23} J/K; but when W increases by the same amount from 800 to 1000, S increases by only 0.3×10^{-23} J/K. The following mathematical operations, based on equation, show analytically what Figure shows visually.

Because $W = L^N$, where N = the number of particles in the system and L = the number of options per particle,

$$S = k \ln W = k \ln L^N = Nk \ln L$$

Doubling N gives $S = k \ln L^{2N} = 2Nk \ln L$. Thus S goes up in direct proportion to N. However, doubling L gives $S = k \ln (2L)^N = Nk (\ln 2 + \ln L) = Nk \ln L + Nk \ln 2$. In this case, S goes up incrementally by N ln 2 each time the number of options per particle increases by a factor of 2.

The Boltzmann and Clausius interpretations of entropy are thus in agreement that the more disordered a system is initially, the less difference an incremental change in the number of arrangements makes. If a state, "a", has more possible arrangements than a state, "b", then state "a" is more probable than state "b". The Boltzmann interpretation of entropy explicitly recognizes the connection between entropy and probability that we developed intuitively. According to this interpretation, the second law of thermodynamics says that the universe seeks states of higher probability.

Example: Consider a system consisting of a pair of dice in a closed box. Suppose that we have no information about this state of the system other than that there are 2 dice in a box. How many "particles" constitute this system? How many arrangements are possible for the system in this state? What is the entropy of the system in this state?

Now suppose that the system is known to be in a "state" in which the sum of the numbers showing on the top faces of the two dice is 2. What is the entropy of this state? What is the entropy of the state in which the number total is 7? The most probable state of the system is the one for which we know only that there are two dice in a box. What is the next most probable state (numerical total) of the system? Discuss the entropies of the least and most probable numerical totals (states).

Solution: The first state to be considered is the one described by saying that there are 2 dice in a box. We will consider each die to be a particle. Then the system consists of 2 particles. Each particle can be in any of 6 "locations" (a die can have 1, 2, 3, 4, 5, or 6 showing on its top face). The total number of arrangements of the 2 particles in the described state is $6 \times 6 = 6^2$, or 36. This is the number of arrangements possible for the system so

$$S = k \ln W = 1.38 \times 10^{-23} \text{ J/K} \times \ln (6^2) = 4.95 \times 10^{-23} \text{ J/K}$$

To assess the probabilities and entropies of the other described states (numerical totals of 2 and 7), we must figure out the number of ways to obtain each state:

State (Numerical Total)	*Die 1*	*Die 2*	*# Arrangements*
2	1	1	1
3	1	2	
	2	1	2
4	1	3	
	2	2	
	3	1	3
5	1	4	
	2	3	
	3	2	
	4	1	4
6	1	5	
	2	4	
	3	3	

	4	2	
	5	1	5
7	1	6	
	2	5	
	3	4	
	3	4	
	5	2	
	6	1	6

and so on.

The most probable state is 7, because 6 different arrangements of the system produce it.

The least probable states are 2 and 12, because each is produced by only one arrangement. State 7 has highest probability and highest entropy, with value k ln 6; states 2 and 12 have lowest probability and lowest entropy, with value k ln 1 = 0.

The entropy associated with states (numerical totals) of dice may be thought of as follows. A blindfolded person is told that the dice show a total of 2, and is asked what number is showing on each die. S/he can, with certainty, state that each die shows 1.

In contrast, when told that the dice show a total of 7, s/he can only guess, with 1 chance in 6 of being correct, at the numbers on individual die. S/he is uncertain. Finally, when told only that there are 2 dice in a box, s/he has only 1 chance in 36 of guessing the numbers on individual die. The extent of uncertainty is a measure of the entropy of the state of the system.

ENTROPY CHANGES IN CHEMICAL REACTIONS

We now turn to the problem of entropy changes in chemical reactions. To calculate these will require that we have available quantitative data on the entropy of 1 mole of each pure substance participating in the reaction. Such data are obtained experimentally by judicious combination of the Boltzmann and Clausius interpretations of entropy. Before looking at things quantitatively, though, it is best to develop a qualitative feel for entropy changes in processes. That is where we begin.

Qualitative Considerations

Is it possible to predict simply by examining the equation for a physical or chemical process whether or not the process is spontaneous from left to right? The answer: sometimes. In order to make such a prediction, it is necessary to demonstrate that DS_{univ} for the process is greater than zero. This in turn requires consideration of both DS_{sys} and DS_{surr}: $DS_{univ} = DS_{sys} + DS_{surr}$

If we can demonstrate that $DS_{univ} > 0$ for a process of interest, we may conclude that the process will spontaneously occur. If instead we demonstrate that $DS_{univ} < 0$ for the process, we may conclude that the process is not spontaneous, but that the reverse process is. It is important to realise that DS_{univ} may be positive even when either D S_{sys} or DS_{surr} (but not both) is < 0. Only the total of the two contributions matters. Thus entropy may decrease in the system by, say, 100 J/K, as long as it increases in the surroundings by more than 100 J/K, so that DS_{univ} is > 0.

Similarly, S may decrease in the surroundings, as long as it increases more in the system. We have already had quite a lot to say about the entropy of systems consisting of single pure substances, S_{sys}, and the factors affecting it. We must now extend these ideas to processes in which more than one pure substance may be involved. Several guidelines are useful:

Guideline: DS_{sys} for a chemical or physical process is positive when

- The number of moles of gas increases;
- The number of independent particles increases;
- The system volume increases at constant temperature;
- Bonds are broken.

Example: Predict the sign of DS_{sys} for each process, and explain the prediction.

$$CH_4(g) \rightarrow C(g) + 4\,H(g)$$
$$H_2O(g) \rightarrow H_2O(l)$$

Solution: In the process, $CH_4(g) \rightarrow C(g) + 4H(g)$, bonds are broken and no bonds are formed to replace them. Disorder will surely increase in this process, and $DS_{sys} > 0$. Note that both the number of independent particles and the number of moles of gas also increase, reinforcing our prediction. Because bonds are broken in this process, we also predict $DH > 0$; bond breaking is always endothermic. Because this process involves atomization of 1 mole of methane, DH for the process is the enthalpy of atomization of methane.

In the process $H_2O(g) \rightarrow H_2O(l)$, a highly disordered gas is converted to a relatively organized liquid. Disorder decreases, and $DS_{sys} < 0$. We expect condensation of vapour to be exothermic: $DH < 0$. Energy leaves the system and enters the surroundings in this process.

From the two processes considered in Example, we can develop an important and useful generalization. For a process in which the bonds of a substance are ruptured, both DS and DH are > 0. When bonds are broken, the system climbs out of a potential energy well. This requires that molecular forces be overcome, a process that is always endothermic: $DH > 0$. When forces are overcome, increased particle freedom inevitably results: $DS > 0$.

$$\text{Bonds} \rightarrow \text{no bonds } DH > 0; DS > 0$$

$$\text{Aggregated state} \rightarrow \text{deaggregated state } DH > 0;\ DS > 0$$

When methane is atomized—that is, deaggregated—covalent bonds are ruptured, and the five atoms are free to move independently; when water condenses (aggregates), bonds are formed as intermolecular forces operate in the liquid phase. The freedom of motion of the molecules is more restricted, and their potential energy is lower than in the gas phase.

Example: Predict qualitatively the sign of DS_{sys} and DH_{sys} for the reaction,

$$F_2(g) + H_2(g) \rightarrow 2HF(g)$$

Solution: Here we have little help. This can not be viewed as a process that involves only the breaking or making of bonds. Although F-F and H-H bonds must be broken to form fluorine and hydrogen atoms, they are replaced with

H-F bonds in the products. The moles of gas stays the same in the process (2 moles in the reactants replaced by 2 moles in the products), as does the number of free particles. It is reasonable in such a case to predict that the entropy change will be small: DS_{sys} approximately 0. However, the enthalpy change depends on the relative bond strengths. We cannot qualitatively predict this without considerable experience. We conclude:

DS_{sys} approximately 0

DH_{sys} cannot predict

Example conveys another important lesson. In most chemical reactions, bonds in the reactants are broken; but bonds are formed in the products. In the majority of cases, the number of chemical bonds in reactants and products is the same, because the same atoms, with the same valences, are present. In such processes, the vast majority, it is not possible to use equation to qualitatively assess the entropy and enthalpy changes. In most chemical reactions, the atoms proceed from the potential well of the reactants to the potential well of the products.

Usually the same number of chemical bonds is involved in both states, and a simple qualitative prediction of DS and DH based is not possible. Instead, the overall enthalpy change depends upon which of the two potential wells is deeper; and the overall entropy change depends on whether the reactants or the products are more disordered. The process shown in the figure is endothermic; the sign of its entropy change is not discernible from the potential well diagram.

Example: For each reaction, determine the change in the number of covalent bonds when reactants are converted to products.

$$CH_4(g) + 2\,O_2(g) \rightarrow CO_2(g) + 2\,H_2O(l)$$
$$CO(g) + H_2O(g) \rightarrow CO_2(g) + H_2(g)$$
$$2\,SO_2(g) + O_2(g) \rightarrow 2\,SO_3(g)$$
$$4\,NH_3(g) + 5\,O_2(g) \rightarrow 4\,NO(g) + 6\,H_2O(g)$$
$$C_6H_{12}O_6(s) + 6\,O_2(g) \rightarrow 6\,CO_2(g) + 6\,H_2O(l)$$

Solution: In each case, the number of bonds in each reactant or product molecule can be counted from a correct

Lewis structure of the molecule. You should verify each bond number given below, if you have difficulty.

	$CH_4(g)$ +	2 $O_2(g) \rightarrow$	$CO_2(g)$ +	2 $H_2O(l)$
#bonds	4	4	4	4
Change in number of bonds = 0				

	CO(g) +	$H_2O(g) \rightarrow$	$CO_2(g)$ +	$H_2(g)$
#bonds	3	2	4	1
Change in number of bonds = 0				

	2 $SO_2(g)$ +	$O_2(g) \rightarrow$	2 $SO_3(g)$
#bonds	6	2	8
Change in number of bonds = 0			

	4 $NH_3(g)$ +	5 $O_2(g) \rightarrow$	4 NO(g) +	6 $H_2O(g)$
#bonds	12	10	8	12
Change in number of bonds = –2.				

	$C_6H_{12}O_6(s)$ +		6 $O_2(g) \rightarrow$	6 $CO_2(g)$ + 6 $H_2O(l)$
#bonds	24	12	24	12
Change in number of bonds = 0.				

We must now address DS_{surr}. This turns out to be a fairly simple matter. We offer Guideline 7 as a means of assessing DS_{surr}:

Guideline: If heat enters the surroundings during a process, $DS_{surr} > 0$; if heat leaves the surroundings, $DS_{surr} < 0$.

Heat entering the surroundings increases the random motion of molecules; disorder increases. Heat leaving the surroundings decreases the random thermal motion of molecules; disorder decreases. We can make this guideline even more useful. For processes carried out at constant pressure, q = DH_{sys}. The heat entering the surroundings is thus $-DH_{sys}$! If we can make a qualitative assessment of DH_{sys}, we can easily make an assessment of DS_{surr}.

Example: Assess the spontaneity of each process.

$Zn(s) + 2\ HCl(aq) \rightarrow ZnCl_2(aq) + H_2(g)$ $DH_R = -153.9$ kJ

$Ba(OH)_2\ 8H_2O + NH_4NO_3 \rightarrow Ba(NO_3)_2 + 2\ NH_3(aq) + 10\ H_2O$ $DH_R = 62.3$ kJ

$2\ H_2(g) + O_2(g) \rightarrow 2\ H_2O(l)$$DH_R = -572$ kJ

Solution: We qualitatively assess DS_{sys} and DS_{surr} for each process and attempt to draw a conclusion about the sign of DS_{univ}.

$Zn(s) + 2\ HCl(aq)$ — $ZnCl_2(aq) + H_2(g)$ $DH_R = -153.9$ kJ

In this process, gas is produced from reactants in condensed phases. Thus $DS_{sys} > 0$. In addition, the reaction is exothermic, meaning that heat passes into the surroundings. Thus $DS_{surr} > 0$. We conclude that DS_{univ} must be greater than zero; Zn(s) spontaneously reacts with aqueous HCl to give the indicated products.

$Ba(OH)_2\ 8H_2O + NH_4NO_3 \rightarrow Ba(NO_3)_2 + 2\ NH_3(aq) + 10\ H_2O$ $DH_R = 62.3$ kJ

In this process, two solids are converted to a third solid, 2 moles of dissolved gas, and 10 moles of liquid water. Thus water molecules previously confined to the crystal lattice of a reactant have enhanced freedom in the products, and two additional water molecules have formed.

Ammonia (NH_3) that is confined in an ionic lattice in the reactants is free to explore the volume of the aqueous solution in the products. It is reasonable to predict that entropy increases: $DS_{sys} > 0$. However, the reaction is endothermic, meaning that heat leaves the surroundings. Thus $DS_{surr} < 0$. We have no information about relative magnitudes, and can say only that if $DS_{sys} > -DS_{surr}$, reaction will be spontaneous.

$2\ H_2(g) + O_2(g) \rightarrow 2\ H_2O(l)$ $DH_R = -572$ kJ

Three moles of gaseous reactants are consumed in this process. We confidently predict $DS_{sys} < 0$. However, reaction is strongly exothermic, so heat enters the surroundings, leading to $DS_{surr} > 0$. Again, without quantitative information, we cannot make a firm prediction. We can say only that if $DS_{surr} > -DS_{sys}$, the process will be spontaneous.

Equation reveals in general terms whether or not processes of particular types are possible. Two reactions being carried out in reaction chambers at constant pressure. The reaction, $R \rightarrow P$, is exothermic; the process is endothermic. Can equation be used to formulate any general statements about the entropy change of the system for these two situations? Consider first the exothermic process. For an exothermic process, DH_{sys} is negative, so heat flows from system to surroundings. Because heat enters the surroundings, $DS_{surr} > 0$. Thus

An exothermic process having $DS_{sys} > 0$ must be spontaneous;

An exothermic process having $DS_{sys} < 0$ is spontaneous as long as $DS_{surr} > -DS_{sys}$.

For the endothermic process, DH_{sys} is positive, so heat flows from surroundings to system during the reaction. Because heat leaves the surroundings, $DS_{surr} < 0$. It follows that An endothermic process can be spontaneous only if

Fe_2O_3(s)87.40

C(s)	5.74
Mg(s)	23.68
MgO(s)	26.94
H_2O(l)	69.91
C_2H_6O(l)	160.7
H_2O(g)	188.83
N_2(g)	191.61
H_2(g)	130.68
CO_2(g)	213.74

Example: Make sense of the entropy data in Table.

Solution: There are a number of important statements to be made about the absolute entropy data in the table.

The symbol for absolute entropy, S^o, is not preceded by the D symbol. The absolute entropy of a substance is the actual (not relative) disorder content of 1 mole of the substance at 298 K. It is possible to know the actual disorder content because of the recognition that a perfect crystal at 0 K has no disorder. There is thus a natural zero, or reference point, for entropy. In contrast, there is no way to know the actual enthalpy content of a substance because there are so many contributions to enthalpy (energy) that cannot be absolutely measured.

We can measure changes in enthalpy, but not absolute values. Thus, for convenience, we reference enthalpies of compounds to those of the elements in their standard states, which are arbitrarily taken as zero. This is analogous to arbitrarily defining gravitational potential energy equal to zero at the earth's surface. It is a zero-point of convenience, but not of reality.

Second, again in contrast to the case for standard enthalpies of formation,

All absolute entropies, including those for the elements, are positive.

This is easy to understand: it is not possible to have less disorder than is present in a perfect crystal at absolute zero, where the disorder is zero (order is perfect).

Third, it is clear from the data that the absolute entropies of solids are small and those of gases are large. The two liquids in the table have entropies that are intermediate in value.

Absolute entropy tends to be small for solids, large for gases.

This is primarily due to the increasing uncertainty in molecular location in progressing from the solid to the gas phase.

Now focus on the solids. Some of the data are puzzling. Although the oxide of iron, Fe_2O_3, has larger absolute entropy than Fe(s), the oxide of magnesium has lower absolute entropy than Mg(s). This is a result of the lattice structures of the solids and is not readily explained at this point; we will not concern ourselves with it. However, we note also that S^o for solid carbon is very much smaller than that for solid magnesium. The reason for this is that atoms in the carbon lattice are bonded covalently. The bonding electrons are restricted to locations between pairs of carbon atoms, and make only a small contribution to the entropy of the solid. In contrast, magnesium is a typical metal. Its valence electrons are free to move over the entirety of the crystal, making a substantial contribution to the absolute entropy.

Metals tend to have relatively large S°, due to electron freedom.

An examination of the two liquids in the table shows that the absolute entropy for ethyl alcohol is substantially larger than that for water. In fact, the value for ethyl alcohol is not much less than that for gases. This is attributed to the molecular complexity of ethyl alcohol, which has 9 atoms per molecule, compared with 3 atoms per molecule of water.

Absolute entropy increases with molecular complexity.

This generalization is reinforced when we examine the gases. Of the gases listed, the least complex, H_2, has the smallest S°, and the most complex, CO_2, has the largest value. But what about H_2 and N_2, which are both diatomic yet differ substantially in S°? Here the difference in complexity originates not in the number of atoms per molecule, but in the number of electrons per atom. Generally, it is found that as the number of electrons per molecule increases, so does S°. The number of electrons increases as molar mass increases, leading to our last generalization. Absolute entropy increases with molar mass.

QUANTITATIVE ENTROPY CHANGES IN CHEMICAL REACTIONS

If we know the absolute entropy for each substance in a chemical reaction, it is easy to calculate the entropy change, DS^o_R, for the reaction. We find the sum of the absolute entropies of the products and subtract from it the sum of the absolute entropies of the reactants. Of course, if a reaction involves more than one mole of a substance, we must multiply its absolute entropy by this number of moles. The entropy change so obtained is automatically a standard entropy change because the absolute entropies are standard values.

That is, the absolute entropy is the entropy of 1 mole of a substance in its standard state at 298 K. In equation form: $DS^o_R = S[S^o \text{(products)}]–S[S^o \text{(reactants)}]$

Example: Predict the sign of DS^o_R for the reaction of Mg(s) with CO_2(g), then calculate its value.

$$2Mg(s) + CO_2(g) \rightarrow 2MgO(s) + C(s)$$

Solution: Our prediction is that the entropy of the solid products will be lower than that of the reactants, one of which is a gas with presumably large S^o. Thus DS^o_R is expected to be < 0. Now we calculate it, using data from Table :

$$DS^o_R = S[S^o(\text{prod})]–S[S^o(\text{react})]$$
$$= (2)(26.94) + 5.74–(2)(32.68)–213.74 \text{ J/K}$$
$$= –219.48 \text{ J/K}$$

The system becomes more ordered as a result of reaction, a result that we qualitatively predicted.

Based on the calculated value of DS^o_R, would we expect the reaction to spontaneously occur? The initial response might be "no, because the disorder in the system decreases." However, we can test this prediction by trying the reaction. A block of dry ice (solid CO_2) is cut in half. A cup-shaped depression is carved out in the centre of the bottom half and filled with magnesium turnings.

The magnesium is then heated with a propane torch until it is hot enough to start to burn. At this point, the top half of the dry ice block is replaced, covering the cup. The reaction begins slowly, but then escalates until the dry ice block glows white from the light emitted by the magnesium as it reacts violently with the CO_2 vapour produced in the cup, and a white smoke of MgO spews forth from the crack between blocks. After watching this spectacular display, we conclude that the reaction is indeed spontaneous, despite the prediction based on the dramatically negative value of DS^o_R for the reaction.

The problem with our prediction is that it ignores the entropy change of the surroundings. A Second-Law prediction of spontaneity (or lack thereof) must be based on DS for the universe, which includes not only DS_{sys}, but also DS_{surr}. To rationalize the spontaneity of the Mg-CO_2

reaction, we must calculate DS_{surr} in addition to DS_{sys}. We can do this with the help of one more guideline:

Guideline 8: The entropy change for the surroundings is calculated by dividing the heat put into the surroundings by the temperature of the surroundings:

$$DS_{surr} = q_{surr}/T_{surr}$$

But the heat put into the surroundings is the negative of the heat of reaction, q_R. Thus$DS_{surr} = -q_R/T_{surr}$

The reaction is done at constant pressure, because it is open to the atmosphere. Thus $q_R = DH_R$. Finally, then

$$DS_{surr} = -DH_R/T_{surr}$$

The other two substances are elements in their standard states, and therefore have $DH_f^o = 0$.

$$DH^o_R = 2\ DH^o_f(MgO) - 1\ DH^o_f(CO_2)$$
$$= 2(-601.7)-(-393.5)\ kJ$$
$$= -809.9\ kJ$$

Then $DS_{surr} = 809900\ J/298\ K = 2718\ J/K$

and $DS_{univ} = 2718 - 219.5 = 2498\ J/K$

When the entropy change in its entirety is considered, the reaction is hugely spontaneous.

Standard Entropy of Atomization

The standard entropy of atomization of a substance, $DS^o{}_{atom}$, is the entropy change that accompanies the conversion of 1 mole of the substance in its standard state to its constituent atoms in the gas phase at a specified temperature, usually 298 K. Several atomization processes and their corresponding standard entropy changes are shown below:

$$CH_4(g) \rightarrow C(g) + 4H(g)\ DS^o{}_{atom} = 430.7\ J/K$$
$$H_2O(l) \rightarrow 2H(g) + O(g)\ DS^o{}_{atom} = 320.6$$
$$C_6H_{12}O_6(s) \rightarrow 6C(g) + 12H(g) + 6O(g)\ DS^o{}_{atom} =$$

We can make several observations about standard entropies of atomization. First, they are entropy changes, because they characterize processes. Thus they must include the symbol D, in contrast to absolute entropies. Second, they are always positive. Bond breaking always increases disorder, because atoms that were previously bound together can now move independently. The standard entropy (change) for a chemical reaction of any type can be calculated from the standard entropies of atomization of the reactants and products by imagining a Hess's Law cycle in which, first, the reactants are atomized, requiring the input of the total of the standard entropies of atomization of the reactants; and second, the atoms are recombined to form products, with the release of the total of the standard entropies of atomization of the products. We have used this Hess's Law view of a chemical process several times before, and it should be familiar to you now. The overall change in entropy for the reaction, then, is given by equation:

$$DS_R^o = S[DS^o_{atom}(reactants)]–S[DS^o_{atom}(products)]$$

Standard entropies of atomization provide an alternative to absolute entropies for the determination of entropy changes in chemical reactions.

THE GIBBS FUNCTION

Example suggests that, at constant temperature and pressure,

$$DS_{univ} = DS_{sys}–DH_{sys}/T$$

The constant pressure restriction is necessary to use DH_{sys} in place of q; the constant temperature restriction is necessary to use $DS_{surr} = q_{surr}/T$, which is an isothermal expression. Equation is interesting because it expresses the entropy change of the universe solely in terms of system functions! To put both sides of the equation into energy units, we rewrite it as

$$TDS_{univ} = TDS_{sys}–DH_{sys}$$

This suggests that we define a new energy function of the system, usually called the Gibbs Free Energy and symbolized G, such that: G = H–TS

A change in G at constant temperature is then given by: DG = DH–TDS

and it is clear that DG, the change in a system function, measures TDS_{univ}. This new function, G, incorporates the influences of both enthalpy and entropy in determining the spontaneity of a process and enables us to reframe the spontaneity criterion of the second law of thermodynamics: At constant T and P, processes having DG < 0 are spontaneous; they tend to proceed left to right as written, favoring the final state. Processes having DG > 0 are non-spontaneous (impossible); they tend to proceed right to left as written, favoring the initial state.

The change in Gibbs Free Energy, DG, reflects a compromise in the effects of DH and DS on the spontaneity of a process. For the atomization of methane, bonds are broken in converting reactants to products. Because bond breaking is always endothermic, yet leads to increased freedom of motion of the atoms involved, both DH and DS for the process are positive. Reactants are favored by enthalpy; but products are favored by entropy. Whether the process is spontaneous or not is determined by the relative magnitudes of DH and TDS in equation. For the atomization of methane at 298 K, the enthalpy "wins"; the process is not spontaneous from left to right.

Example: Apply the new criterion of spontaneity to the Mg-CO_2 reaction of example.

Solution: We use DG = DH–TDS at 298 K. The enthalpy and entropy changes of reaction have already been calculated in the previous example.

DG = DH–TDS = –809.9–(298)(-0.219) kJ
= –745 kJ

The reaction is predicted to be spontaneous. Note that some care is required to express the changes in enthalpy and entropy in the same units. Standard Free Energy of Formation, DG_f^o. Application of the free energy criterion to chemical reactions is easy. One approach is to calculate DH_R^o and DS_R^o from tabulated enthalpies of formation and absolute entropies, then calculate DG_R^o for the reaction using equation. A second approach is to use so-called standard free energies of formation, which have been tabulated by chemists for many important substances. The standard free energy (change) of formation of a substance is the free energy change that occurs when 1 mole of the substance is made from its elements in their standard states. The definition is analogous to that of standard enthalpy (change) of formation. For calcium carbonate, the standard free energy of formation is the free energy change for the following reaction:

$Ca(s) + C(s) + 3/2\ O_2(g) \rightarrow CaCO_3(s)$ [$DG_R{}^o$ is by definition $DG_f{}^o$]

Table gives standard free energies of formation at 298 K for calcium carbonate and a number of other common substances. For an element in its standard state, the standard free energy of formation is zero, because there is no change in either enthalpy or entropy when 1 mole of an element is formed from itself. If we apply Hess's Law to the free energy change in the same way, we arrive at equation:

Table. Standard Free Energy of Formation, kJ/mole

Substance	DG_f^o
$CaCO_3$	–1128.76
CaO	–604.17
$CH_4(g)$	–50.8
$C_2H_2(g)$	209.2
$C_2H_4(g)$	68.11
$C_6H_{12}O_6(s)$	-910.4
$CO_2(g)$	-394.4
$H_2O(l)$	–236.81
MgO(s)	–569.6

$DG_R{}^o = S[DG_f{}^o(prod)] - S[DG_f{}^o(react)]$

We apply this to an important reaction.

Example: Is the photosynthesis reaction spontaneous at 25 °C?

$$6\ CO_2(g) + 6\ H_2O(l) \rightarrow C_6H_{12}O_6(s) + 6\ O_2(g)$$

Solution: To answer the question, we require standard free energies of formation for all participants in the reaction.

Substance	DG_f^o, *kJ/mole*
$CO_2(g)$	–394.4
$H_2O(l)$	–236.8
$C_6H_{12}O_6(s)$	–910.4
$O_2(g)$	0

$$DG_R{}^o = DG_f{}^o\ (C_6H_{12}O_6) - 6DG_f{}^o\ (CO_2) - 6DG_f{}^o\ (H_2O)$$
$$= -910.4 - 6(-394.4) - 6(-236.8) = 2877\ kJ$$

The reaction is predicted to be non-spontaneous

(impossible) at ordinary temperature, yet it occurs on a tremendous scale all around us. An interesting question: why does photosynthesis occur?

Standard Free Energy of Atomization

The standard free energy (change) of atomization of a substance, DG^o_{atom}, is the free energy change accompanying the conversion of 1 mole of the substance in its standard state to atoms in the gas phase. The standard free energy of atomization is related to the standard enthalpy and entropy changes of atomization via equation, which is just applied to the process of atomization.

$$DG^o_{atom} = DH^o_{atom} - TDS^o_{atom}$$

Tabulated values of DG^o_{atom} at T = 298 K. Equation shows how they can be used, according to the usual Hess's Law cycle in which reactants are converted to products via the gas phase atoms, to calculate the overall standard free energy change for any chemical reaction of interest.

$$DG_R{}^o = S[DG^o_{atom}\ (\text{reactants})] - S[DG^o_{atom}\ (\text{products})]$$

This equation can be used as an alternative to equation involving standard free energies of formation.

Because both the standard enthalpy and entropy of atomization are invariably positive numbers, it is clear from that the standard free energy of atomization can be positive or negative, depending on the relative magnitudes of the two terms. It turns out that the enthalpy term dominates at 298 K for most substances, so most of the values are > 0. At high temperature, the second term becomes dominant, so that many standard free energies of atomization become negative.

The Interpretation of DG as "Free" Energy. For a process carried out at constant temperature and pressure, the value of DG provides a measure of the portion of the total energy produced by the process that is "free" to be used for doing work of some kind. This is the origin of the name, free energy. To clarify this, we consider the reaction of hydrogen and oxygen to form water at 298 K.

$$2H_2(g) + O_2(g) \rightarrow 2\ H_2O(l)$$

By now, we are aware that this process is exothermic,

and can use standard enthalpies of formation or atomization to show that DH_R^o = –572 kJ. It is reasonable to wonder whether all of this released energy can be harnessed to do work; for example, to run an engine. In fact, only a portion of this total released energy can be so used, for reasons that we now consider.

A qualitative assessment of the entropy change for the reaction reveals that $DS_{sys} < 0$. Using absolute entropies, it is a simple matter to show that DS_R^o = –326 J/K. For the reaction to be spontaneous, it is necessary that DS_{univ} be positive; thus the entropy of the surroundings must increase by a minimum of 326 J/K to guarantee spontaneity. At 298 K, this increase in entropy of the surroundings requires that at least TDS_{surr} = 97.1 kJ of heat be provided to the surroundings during the reaction.

This heat must be drawn from the total energy produced by the reaction.

As a consequence, only 475 kJ of the total 572 kJ is actually available to do work. The remainder must be paid to the surroundings as heat to guarantee an overall entropy increase.

In contrast, for an exothermic reaction having $DS_{sys} >$ 0, it is not necessary for any energy to be transferred to the surroundings to guarantee that $DS_{univ} > 0$. For such a process, not only is the whole of DH free to do work; but the quantity T DS_{sys} is also available. Work in excess of the magnitude of DH is obtainable from such a process.

The "free energy" available from processes having varying signs and magnitudes of DH_{sys} and DS_{sys}.

The Temperature Dependence of DG

Experiment shows that neither DH nor DS for chemical reactions is particularly sensitive to temperature. Thus a reaction run at 400 K is expected to have nearly the same standard enthalpy and entropy changes as the standard values at 298 K.

Both DH_R^o and DS_R^o can therefore be treated as constants within a fairly large range of temperature centered on 298 K. Since $DG^o = DH^o – TDS^o$, it follows that a plot of $DG_R{}^o$ against T should be linear with

$$\text{Slope} = -DS_R{}^o$$

$$\text{Intercept} = DH_R{}^o$$

Example: Consider the simple physical process

$$H_2O(l) \rightarrow H_2O(g) \text{ 1 atm}$$

for which $DH_R{}^o > 0$ and $DS_R{}^o > 0$. Predict the general form of a plot of $DG_R{}^o$ versus T for the process, and discuss significant aspects of the plot.

Solution: Since the standard enthalpy and entropy for the process are both positive, the plot should have a positive intercept and a negative slope. We make the following observations from the plot:

- At low T, the TDS term is small, so the DH term dominates; DG is positive;
- At high T, the TDS term is large, and dominates; DG is negative;
- At the point at which the plot crosses the T axis, the enthalpy and T-entropy terms are balanced. At the crossover T, DG = 0, meaning that the enthalpy and entropy changes for vaporization of water to form water vapour at 1 bar are balanced at this temperature. This is of course the normal boiling point, 373 K.

Guideline 9: As T increases, all processes proceed spontaneously (become more favorable) in the direction that increases S.

We have previously stated that the atomization of methane is non-spontaneous at 298 K because the entropy increase is not sufficient to offset the large enthalpy increase that accompanies the rupture of the carbon-hydrogen covalent bonds.

$$CH_4(g) \rightarrow C(g) + 4H(g) \text{ DH, DS} > 0$$

Guideline 9 suggests, however, than an increase in temperature should favour atomization. Indeed, at very high temperature, such that TDS is more positive than DH, atomization becomes spontaneous. Increased kinetic energy (temperature) favors deaggregation.

Example: The Thermodynamics of a Rubber Band. We carry out a simple experiment with a rubber band, which produces a surprising result. A rubber band is stretched and allowed to come to air temperature. It is then allowed to spontaneously contract. Its temperature is measured immediately after the process and is found to be lower than the initial temperature; i.e., contraction of the rubber band is endothermic. Why is contraction endothermic? Why does the rubber band spontaneously contract?

Solution: We apply the First Law to the process. First, since the rubber band contracts very quickly, there is no time for heat flow from the surroundings into the rubber band. Therefore

$$q = 0$$

Second, since the rubber band contracts by itself with no assistance from outside, there is no work done. Thus

$$w = 0$$

The inevitable conclusion from the first law is that

$$DE = 0$$

The internal energy of the rubber band does not change when it contracts.

That the temperature of the rubber band decreases on contraction implies that the average kinetic energy of the rubber molecules decreases:

$$DKE < 0$$

Since $DE = DKE + DPE$, it follows that $DPE > 0$. When the rubber band contracts, its potential energy increases in some way. Realise that this conclusion is forced on us by experiment (the observed decrease in T) and application of the First Law. It is possible to reach this conclusion without saying anything about the molecular nature of the rubber band. It is a conclusion arrived at by macroscopic observation. To go further—to understand the manner in which potential energy increases for the rubber band—requires that we switch to the molecular perspective.

Rubber is a polymer, which means that its molecules are like long strings. When the rubber band is stretched, so are the molecules. They align in the direction of stretching and thereby approach each other closely along their entire lengths. The intermolecular forces are strong, and PE is correspondingly low. When the rubber band contracts, the molecules relax and become tangled, like a mass of spaghetti, at right. In this state, molecules are in contact only where they cross over each other.

Intermolecular forces are weak, and PE is correspondingly high. According to this simple picture of the rubber molecules, intermolecular PE increases when the rubber band contracts. Since there is no time for heat to flow into the rubber band to supply the energy needed to overcome intermolecular forces, the rubber band gets the energy from itself; it converts molecular kinetic energy to potential energy, and cools.

Example: The spokes of a bicycle wheel are replaced by rubber bands, and the wheel is suspended above the ground as shown, supported by the axle. Initially the wheel is stationary (i.e., not rotating). A bank of heating lamps is placed behind the wheel, close to the rubber bands. When the lamps are turned on, which way will the wheel rotate?

Solution: The wheel will rotate counterclockwise. See if you can develop a reasonable explanation.

The qualitative and quantitative aspects of entropy and have applied the ideas to chemical and physical reactions. In the course of this development, entropy has emerged as an odd quantity. It is a state function for the system, like energy; yet it is very different from energy.

When something happens in the world, whether it is a natural process like a volcanic eruption or a man-driven process like the manufacture of an automobile, energy may be changed in form, but it is not changed in amount: the total energy of the universe is constant (conserved). However, this is not true of entropy.

Entropy is created every time something happens! In fact, only events that create entropy are allowed. This gives a preferred direction — a one-wayness — to events that has led to the idea that entropy is in some way connected to the unidirectional flow of time. Further, the fact that entropy is continually being created means that it is not a conserved quantity for the universe. Yet we have defined it to be conserved for the system of interest. Therefore, it is not conserved for the surroundings.

This is a very odd state function indeed. It is conserved for what we choose to define as the system, but it is not in general conserved outside of that system. The justification for dealing with entropy in this strange way is that it works in both the explanatory and the predictive senses. It is therefore very important to grapple and come to terms with the strangeness of entropy.

Supplement: But What About When DG

We have seen that when DGº < 0 for a chemical or physical process, the process is spontaneous left to right from a starting point in which all reactants and products are in their standard states. When DGº > 0, the process is spontaneous right to left from a starting point in which all reactants and products are in their standard states. It is logical, then, that when DGº = 0, the process is equally spontaneous in both directions; it is in equilibrium under standard conditions. In this situation, reactants and products exist together indefinitely, each in its standard state, because the driving force for conversion of reactants to products is balanced by the driving force for conversion of products to reactants.

We learned that this situation is dynamic. At equilibrium, both conversions continue to take place at the molecular level. They take place at equal rates, however, so that from the macroscopic perspective, amounts of reactants and products do not change. The conversion of liquid water to water vapour is a process with which everyone has some familiarity, so we will use this process to explore the connection between equilibrium and DGº = 0. The process is shown below:

$$H_2O(l) \text{ Ÿ } H_2O(g)$$

We have used the double arrow to indicate that the process establishes dynamic equilibrium in a closed container. Suppose we are interested in determining the temperature at which liquid water is in equilibrium with water vapour at 1 atm pressure (i.e., the temperature at which the vapour pressure of water is 1 atm). We can estimate this temperature by taking advantage of the fact that DG° = 0 at this temperature. It can be shown that $DH_R°$ = 44.05 kJ and $DS_R°$ = 118.7 J/K for this process. At any particular temperature, it is true that $DG_R° = DH_R° – TDS_R°$. Setting $DG_R° = 0$ and solving for T gives: $T = DH_R°/DS_R°$

Thus the temperature at which liquid water exists in equilibrium with water vapour at 1 atm pressure is 370 K. At this temperature, the conversion of liquid to vapour and vapour to liquid are equally spontaneous. We recognize that 370 K is the normal boiling point of water (actually, we do not obtain exactly 373 K because we have used values of the standard enthalpy and entropy for the process that are valid at 298 K. The values are somewhat different at 373 K). For a pure substance, the temperature at which DG° for conversion of liquid to vapour = 0 is the normal boiling point.

Chemical reactions carried out in closed systems also come to dynamic equilibrium, in which the forward reaction rate (conversion of reactants to products) and the reverse reaction rate (conversion of products to reactants) are equal. For a chemical reaction, $DG_R° = 0$ implies that reactants and products exist in equilibrium with each substance in its standard state. The temperature at which this situation exists can be estimated using the same approach used for phase equilibrium above. For example, reaction has $DH_R°$ = 57.2 kJ and $DS_R°$ = 175.7 J/K. Note that these values are consistent with bond breaking.

$$N_2O_4(g) \rightarrow 2NO_2(g)$$

$DG_R° = 0$ at a temperature such that $T = DH_R°/DS_R°$, or 326 K. At this temperature, we can have N_2O_4 and NO_2 in the same container, each with a partial pressure of 1 bar, with forward and reverse reaction tendencies exactly the same. In other words, the reaction is at equilibrium.

A very interesting aspect of equation deserves mention. Unless $DH_R°$ and $DS_R°$ have the same sign (both negative or both positive), equation yields a negative Kelvin temperature! The physical significance of this is that processes for which $DH_R{}^o$ and $DS_R{}^o$ have opposite signs can not have $DG_R{}^o = 0$ at any temperature. They are either spontaneous under standard conditions at all temperatures (exothermic with positive entropy change) or non-spontaneous under standard conditions at all temperatures (endothermic with negative entropy change).

Our study of phase equilibrium revealed that two phases of a pure substance can be in equilibrium over a fairly wide range of temperature and pressure conditions. At 298 K, liquid water is in equilibrium with water vapour at 23.8 torr (0.0313 atm) in a closed container.

$$H_2O(l) \rightarrow H_2O(g)$$

P = 0.0313 atm

This process is equally spontaneous in both directions; consequently it must be true that DG = 0. Because the pressure of water vapour is not 1 atm, however, the water vapour is not in its standard state, and DG is not a standard value. It is written without the superscript o to indicate this. The standard free energy change for process can be calculated in the usual way: $DG_R{}^o = DH_R{}^o - TDS_R{}^o = 8.7$ kJ. This tells us that the process in which water vapour at 1 atm is formed from liquid water at 25°C is nonspontaneous; steam at 1 atm pressure spontaneously condenses at 25°C. Similarly, reaction establishes equilibrium at temperatures other than 326 K. In general, the pressures of NO_2 and N_2O_4 are non-standard values, though, unless T = 326 K. For an equilibrium situation at all temperatures other than this, $DG_R = 0$, but $DG_R{}^o$ not equal to 0.

Index

K

L

M

T

U

V

W

X